CHEMICAL DEMONSTRATIONS

for High School and General Chemistry

Over 40 experiments to ignite curiosity and deepen understanding in chemistry classrooms.

Thomas S. Kuntzleman

innovativeink
PUBLISHING
A Division of Kendall Hunt

Cover image Thomas S. Kuntzleman

innovativeink PUBLISHING
A Division of Kendall Hunt

www.innovativeinkpublishing.com
Send all inquiries to:
4050 Westmark Drive
Dubuque, IA 52004-1840

Print ISBN: 979-8-3851-9291-5
Ebook ISBN: 979-8-3851-9292-2

Published in the United States of America

To Jacob, John, and Jackson

—who taught me that scientific observation is best seen through the eyes of a child

Contents

Preface

I have been presenting chemistry demonstrations for students of all ages for over 30 years. I do this for a variety of reasons, but mostly because doing so provides both learners and instructors with benefits that are positive and profound. Describing these impacts in a fair and thorough manner is difficult. However, an experience I had several years ago captures much of the essence of what happens when we do chemistry experiments with and for others.

On rare occasions, the planet Venus passes between the Earth and the Sun. When it does so, it blocks the Sun's rays and therefore appears to observers on Earth as a black dot that slowly moves across the face of the Sun. The passage is analogous to a solar eclipse of the Sun by the Moon, except that Venus is too far away to fully obscure the Sun. This astronomical event is known as the "transit of Venus" and my son, Jacob, and I attempted to witness this rare event with a small telescope in 2012. Unfortunately, viewing the Sun through a telescope can quickly damage your eyes, and we didn't have a protective viewing filter. After some online searching, we learned that we could project the Sun's rays through the telescope onto a piece of paper for viewing. When we did this, we could clearly see a small black disk on the image of the Sun (Figure 1, left). We were simultaneously observing Venus and the Sun! Over time, we watched the black disk slowly move across the Sun's image as Venus made its transit (Figure 1, middle). Just for fun, we decided to see if we could project the image onto Jacob's t-shirt (Figure 1, right).

FIGURE 1: Left: Venus is visible as a black dot just below the top of the image of the Sun. Middle: Over time, the image of Venus moved across the image of the Sun. Right: we successfully projected the image onto Jacob's shirt. All pictures taken with a flip-top phone.

As we experimented together, Jacob and I wondered about the Sun and Venus. What powers the Sun? Why does it give off light? Did you know that Venus has an atmosphere that is made of almost entirely CO_2—and that it has clouds made of droplets of sulfuric acid? How far away is the Sun from Earth? What about Venus? In my experience, observing simple chemistry experiments imbues my students with the same kind of wonder: What do the nucleation sites look like on Mentos candies, and how do they produce bubbles in Diet Coke? Is drinking heavy water bad for you? What happens if we use a different color of food dye in the density bottle?

As the Sun began to set our view became obscured by some trees. We were having such a good time that we decided to find higher ground so we could continue watching. We walked down the neighborhood street and found a spot in Mr. and Mrs. B's backyard where we could more clearly see the skyline (we live in a close-knit neighborhood, so I knew that Mr. and Mrs. B wouldn't mind). We set up the telescope and continued our observations. Not long thereafter Mr. and Mrs. B came outside to ask what we were doing. "We're watching Venus pass in front of the Sun," I replied. "Take a look at Jacob's shirt." As Mr. and Mrs. B viewed the image, we discussed what was causing the large bright circle and tiny black dot. We talked about some facts related to the Sun and Venus, and then they asked how Jacob was doing in school. I asked Mr. B. if he was back to golfing again, and how things were going in Mrs. B's first grade classroom. The Sun set, we said our goodbyes, and Jacob and I headed home.

As we walked home, I wondered how it could be that the simple image of a large and bright circle of light slightly obscured by a tiny black dot was so interesting to Mr. and Mrs. B. After all, they could have gone online to find strikingly beautiful images of both the Sun and Venus at incredibly high resolution captured by NASA scientists. For that matter, Jacob and I could have done the same thing. And yet, Mr. and Mrs. B spent several minutes—and Jacob and I spent the whole afternoon until sundown—monitoring the curious bright and dark circles. It seemed to me our active participation in capturing the images and experimenting in real time both contributed to the fun. That Jacob and I spent time together made the afternoon even more enjoyable.

This special memory of mine provides several insights into the value of performing live chemistry experiments for students. First, chemistry demonstrations engender a sense of *participation*. This is especially true when I ask for volunteers from the class to help me carry out procedures, or when I ask students to predict "what do you think will happen when …" prior to carrying out an experiment.[1] Fortunately, the very nature of scientific endeavors requires participation: gathering materials, setting them up, carrying out experimental steps, and carefully observing outcomes. These acts imbue experimenters and observers with a sense of ownership in the activities, which contributes to greater satisfaction. As alluded to before, Jacob and I could easily have looked online for intricate pictures of Venus crossing the Sun, but we wanted to capture the event for ourselves. We spent hours doing so and unintentionally enticed our neighbors to join in. Unsurprisingly, the educational literature is full of studies that provide evidence that hands-on experimentation increases student success in and appreciation of science. For example, individuals engaged in learning scientific concepts in a hands-on manner have brain scans that show increased activity relative to students involved in more passive methods.[2]

Second, many of the demonstrations I conduct for my students include phenomena they have never seen before. Thus, for many students these demonstrations involve new experiences, which evoke a sense of *anticipation*. Watching the transit of Venus was something neither Jacob nor I had observed before, and we instinctively sought out this new experience. Just like the feeling you get when the curtain is drawn and the lights go down in the movie theatre, or as a roller coaster climbs the hill before its first descent, carrying out a brand-new experiment arouses feelings of expectant wonder. Will The Black Panther defeat the villain in the movie? What will I feel like when the roller coaster is completely upside down during a loop-de-loop? What will happen if I mix Diet Coke and Mentos under a layer of oil? We humans often seek out new adventures because we enjoy the feeling of anticipation. Notice that seeking out new experiences is closely related to *exploration* and *curiosity*, two other aspects of why I think live chemistry demonstrations are so important. Neuroscientists have learned that a sense of curiosity not only enhances learning but also makes it more enjoyable.[3] Mr. and Mrs. B must have been curious when they saw Jacob and I playing around with a telescope in their backyard in the late afternoon. Consequently, they came outside to see what was going on. When we told them about what we were exploring, they were eager to join in. Yes, we humans love to learn something new or to find things out.

One of the things I find most enjoyable about doing science experiments is that they don't have to be complicated to inspire curiosity. Outside of a telescope, Jacob and I only needed a piece of paper (or white shirt) to conduct our experiments on the transit of Venus. How easy is that? As another example, consider what happens when M&M's candies are arranged in a pattern on a plate, and then water is poured into the plate. The dyes from the candy spread out into the water, and yet they don't mix (Figure 2, left). Some strikingly beautiful patterns can be created from this very simple experiment (Figure 2, right). When I first saw this experiment several years ago, I was dumbfounded.[4] How in the world can it be that the colors don't mix? This initial curiosity remained with me for years and has inspired me to carry out several investigations to uncover what might be responsible for the nonmixing effect.[5]

FIGURE 2: Water added to M&M's candies arranged in a pattern on a plate displays beautiful patterns.

Live chemistry demonstrations also promote *learning*, which is the acquisition of new skill or new knowledge (there's that word *new* again). Who doesn't love to discover something or to learn something new? The Nobel Prize-winning physicist Richard Feynman went so far as to state "if you're not having fun, you're not learning. There's a pleasure in finding things out." As Jacob and I learned the new skill of projecting an image of the Sun through a telescope, we shared with each other facts about the Sun, Venus, and some of the other planets.

We both gained new knowledge from each other as we talked. These facts came up in our conversation with Mr. and Mrs. B, too. These discussions flowed as a natural extension of the experiments we engaged in together. For example, we discussed sunspots, which are slightly cooler regions on the Sun's surface. Sunspots show up as dark regions on the Sun's surface. Take a closer look at the image of the Sun projected onto Jacob's shirt (Figure 3). Can you see some small black smudges on the bright disk of the Sun in addition to the black circle caused by Venus? These are sunspots, and Jacob and I unexpectedly noticed these during our observation of the Venus transit.

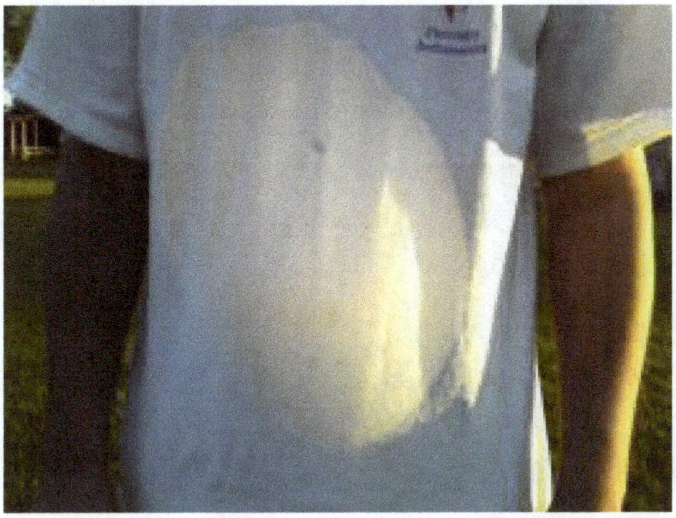

FIGURE 3: Blown-up image of the photo shown in Figure 1, right.

Our unexpected observation of sunspots illustrates yet another benefit of active experimentation: because it very often produces *unexpected results*—especially if you look carefully enough. Remember my amazement at the surprising, nonmixing effect in the M&M experiment? I recently asked Jacob, who is now a physicist, what he most liked about science. Almost without pausing, he said, "I like the surprising nature of science. You often see, learn, and experience things that you never expected." Indeed, when we train ourselves to observe carefully, we find that unexpected results are a common feature of scientific experimentation.

Sometimes the results of science experiments are so surprising that they are accompanied by *humor*. If you are familiar with the chemistry experiment known as "Elephant Toothpaste" (Figure 4), you have an idea of what I'm talking about. During the Elephant Toothpaste reaction, a solution containing hydrogen peroxide is decomposed into water and oxygen gas when a catalyst is added to it:[6]

$$H_2O_2(aq) \rightarrow 2\ H_2O(l) + O_2(g) \text{ (in the presence of a catalyst)}$$

The reaction is usually carried out in a ½ L cylinder that contains some detergent. The detergent makes it so that when the oxygen gas is produced, an enormous amount of foam is generated (Figure 4). When I first saw this reaction, I initially observed a small amount of foam gently pour over the sides of the container (Figure 4, middle panels). Imagine my surprise when, after a slight pause, the foam proceeded to expand rapidly to a total volume of about 10 liters (Figure 4, right). I burst out in laughter at this unexpected result. (For a much less dramatic example, Jacob and I did enjoy a small chuckle when we successfully projected the image of Venus and the Sun onto his shirt).

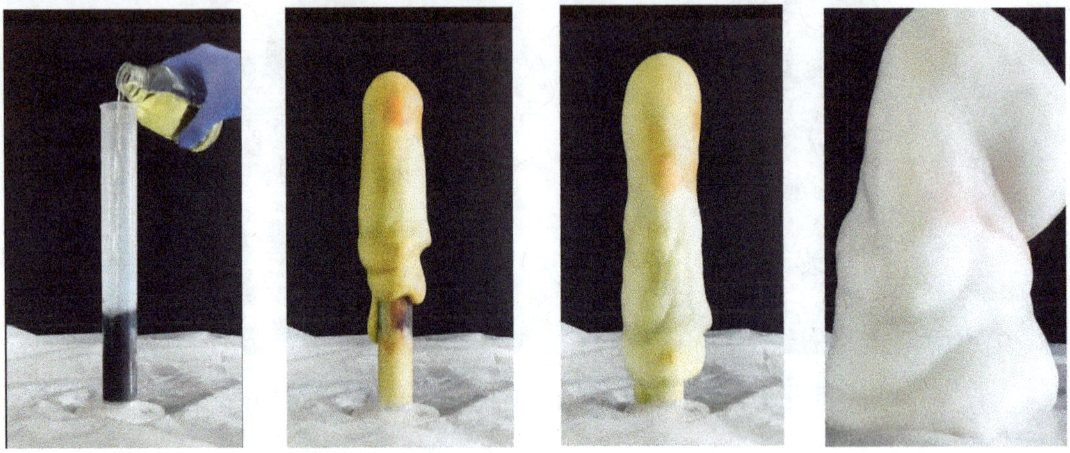

FIGURE 4: The Elephant Toothpaste reaction. Left: A solution containing a catalyst is added to a mixture of hydrogen peroxide and detergent in water. Middle two panels: The resulting chemical reaction produces oxygen gas. Right: The resulting oxygen gas and detergent produce a lot of foam.

Finally, chemistry demonstrations fill us with *wonder*. Many of us recognize *awe* as the inspiration we get when we look at the stars on a clear night or gaze into the eyes of a newborn baby. I happen to feel this emotion in the springtime at night when I sit outside and listen to the loud and resonant chirping of frogs in the swamp behind my backyard. Viewing the transit of Venus allowed Jacob and I to reflect on the vastness of our solar system, and to connect with the enormity of the universe in a small way. The amazement that Jacob and I felt the day we watched the transit of Venus has remained with us as a steadfast companion over the years. This persistent curiosity has continued to motivate us to observe other objects in our solar system and beyond. We have used the telescope to see the Moon; the phases of Venus, Mars, and Jupiter; the Galilean moons; and Saturn and its rings. We think we may have even observed polar ice caps on Mars and Titan, one of the moons of Saturn. We've checked out various comets when they become visible, and Jacob taught me how to locate the Andromeda Galaxy. It's quite easy to have deep and meaningful conversations while pondering the fact that you are observing light that is just reaching your eye after leaving Andromeda 2.5 million years ago. Our common interest in observing the nighttime sky has branched out to other family members, friends, and people in our neighborhood. When we're in the front yard looking at Saturn through the telescope, we'll often coax neighbors who are walking by to come and see. We commonly invite people over when we know we'll have the telescope out for the evening.

These same qualities—and more—are what make chemistry demonstrations so powerful. They allow students to actively engage with matter and its transformations—not through high-resolution pictures or abstract theory, but through direct observation and participation. Soda jetting out of a bottle of Diet Coke, bubbles suspended in mid-air, the entire rainbow of colors flashing through a flask in just a few seconds: These moments touch our souls as well as our senses. They give us all a chance to say, *I have seen it for myself, and I have even understood what is happening.* And in these moments, chemistry becomes not just something to learn, but something to love.

With all this in mind, I present the demonstrations in this book. I hope you find them to be useful, accessible, and intriguing. I also hope you see something new as you carefully observe what's going on or tweak the conditions and see a surprising result. Most of all, I hope these demonstrations help connect your students to a deeper understanding and appreciation of chemistry, and you more deeply to your students.

Happy Experimenting!

REFERENCES

1. Crouch, C. H., Fagan, A. P., Callan, J. P., & Mazur, E. (2004). Classroom demonstrations: Learning tools or entertainment? *American Journal of Physics, 72*(6), 835–838.
2. Kontra, C., Lyons, D. J., Fischer, S. M., & Beilock, S. L. (2015). Physical Experience Enhances Science Learning. *Psychological Science, 26*(6), 737–749. https://journals.sagepub.com/doi/abs/10.1177/0956797615569355

3. Gruber, M. J., Gelman, B. D., & Ranganath, C. (2014). States of curiosity modulate hippocampus-dependent learning via the dopaminergic circuit. *Neuron, 84*(2), 486–496. https://www.cell.com/neuron/fulltext/S0896-6273(14)00804-6
4. Kuntzleman, T. (2013). Fun with M & M's. *Chemical Education Xchange.* https://www.chemedx.org/blog/fun-m-ms
5. Kuntzleman, T. (2021). How does the rainbow candy experiment work? *Chemical Education Xchange.* https://www.chemedx.org/blog/how-does-rainbow-candy-experiment-work
6. Shakhashiri, B. Z. (1983). In *Chemical demonstrations: A handbook for teachers of chemistry* (Vol. 1, pp. 180–185). The University of Wisconsin Press.

Introduction

Chemistry demonstrations have several positive impacts on student learning. This fact is well established in the educational literature—and, in my opinion, is also common sense. After all, chemistry is the study of matter, so it stands to reason that students should spend time observing matter directly. Chemical demonstrations are perhaps the most efficient, convenient, and focused way to achieve this goal.

This book aims to support chemistry teachers—especially those working in high school, introductory, and general chemistry classrooms—to incorporate demonstrations into their lessons regularly. Each demonstration in this volume has up to three key characteristics designed to make such integration practical and meaningful.

The first characteristic is accessibility. Every demonstration described in this collection is easy to set up and carry out. In my own experience as a chemistry teacher, I find I'm far more likely to conduct demonstrations that use inexpensive, easily obtained materials—items that can be purchased a day or two in advance and prepared just minutes before class. Equipping teachers with a wide variety of accessible, low-prep experiments makes it more likely that students will see chemistry in action on a regular basis. In addition, because it is known that student learning improves when chemical topics are connected to familiar, everyday phenomena, using common materials adds a second layer of educational benefit.

The second feature is curricular alignment. Each demonstration connects directly to topics found in the high school and general chemistry curriculum. This alignment was intentional, as demonstrations have been shown to be more effective when tied to specific learning objectives.

The third feature is quantitative rigor. Where appropriate, the phenomena involved in each demonstration are described in a quantitative manner or connected to the kinds of problems commonly encountered in general chemistry homework. After all, quantitative reasoning is central to scientific understanding.

"When you can measure what you are speaking about, and express it in numbers, you know something about it; when you cannot express it in numbers, your knowledge is of a meager and unsatisfactory kind …."
—Lord Kelvin

While not every demonstration contains all three of these features, all include at least one—and many include all three.

I have spent over 30 years instructing students as both a science teacher and chemistry professor, and the demonstrations and accompanying discussions in this book have all stemmed from this work. Several demonstrations in this volume are adapted and expanded from blog posts I previously shared on the *Chemical Education Xchange*, an online community and resource platform for chemistry teachers worldwide. It is my hope that this book will serve as a valuable resource for chemistry teachers seeking to incorporate demonstrations into their classrooms in an efficient, impactful, and conceptually rigorous way. The procedures described in this book are intended solely for use by trained chemistry instructors. Anyone conducting these demonstrations does so at their own risk.

1

The Density Bottle

TOPICS: Classification of Matter, Density

Materials:

- Clear, colorless, and narrow bottle capable of holding 500 mL of fluid.
- 240 mL of deionized water.
- 50 g NaCl (table salt; preferably no additives to avoid cloudiness).
- 280 mL of 91% isopropyl alcohol.
- Small flashlight (optional).

Hazards:

Gloves and goggles should be worn to avoid splashing fluid onto skin and in eyes. Alcohol is flammable; keep from sources of ignition. Work in a well-ventilated area to avoid inhalation of alcohol fumes.

Procedure:

1. Dissolve 50 g NaCl in 240 mL deionized water, stirring until fully dissolved.
2. Pour the salt solution into the bottle.
3. Carefully add 280 mL of 91% isopropyl alcohol to the bottle.
4. Cap the bottle securely and mix gently.
5. Allow the bottle to sit undisturbed on a flat surface. Within a few minutes, three distinct layers will form (Figure 1):
   a. **Top:** Alcohol layer (transparent, but some bubbles may be visible).
   b. **Middle:** Emulsion layer (cloudy).
   c. **Bottom:** Saltwater layer (transparent).
6. Optionally, shine a flashlight through the bottle and observe the behavior of light in each layer.

FIGURE 1: Typical appearance of the bottle in this demonstration

Disposal:

The fluid in the bottle may be saved for future demonstrations.

Discussion:

When alcohol and saltwater are combined, a temporary and unstable mixture of these substances forms. Over the course of a few minutes, this transient mixture separates into two phases. Thus, alcohol and saltwater are said to be **immiscible**. The temporary mixture of these two fluids is called an **emulsion,** and emulsions are heterogeneous mixtures.

The layers in this demonstration illustrate three different classifications of matter. The alcohol layer can be considered to be a pure compound, the saltwater is a homogeneous mixture, and the alcohol–salt water emulsion is a heterogeneous mixture. A light shining through each layer will show different behavior. Light passes through the saltwater layer with little scattering, so the light path is not visible. This is because the sodium and chloride ions in the saltwater completely dissociate so that only individual ions and water molecules exist in the solution. Ions and molecules are much too small to scatter visible light appreciably, so the light passes through unscathed. On the other hand, light passing through the emulsion layer is scattered, and its path of light is visible. In this case, alcohol molecules are not completely separated from one another, but rather form tiny droplets within the saltwater. While these droplets are indeed very small, they are large enough that light scatters off these droplets. Thus, shining light through the layers can distinguish between homogeneous and heterogeneous mixtures.

The relative density of each layer may be noted. The most dense fluid (saltwater, $D = 1.12$ g mL^{-1}) sinks to the bottom, while the least dense fluid (alcohol, $D = 0.79$ g mL^{-1}) floats on top. The emulsion has a density that is intermediate between the two, and so it rests in between these two fluids.

REFERENCE

1. Kuntzleman, T. (2017). The density bottle strikes again. *Chemical Education Xchange*. https://www.chemedx.org/blog/density-bottle-strikes-again

2

The Floating Soap Bubble

TOPICS: density, density of gases, ideal gas law

Materials:

- 1 gal of vinegar.
- 1 lb of baking soda.
- Bubble solution with a wand.
- Tea light candle.
- Lighter.
- A large container that is easy to see through. A large clear plastic container or large aquarium works.

Hazards:

Use great caution when lowering the candle into the container. Keep in mind that the heat from the flame can cause burns.

Procedure:

1. Liberally sprinkle baking soda into the bottom of the very large container. Add at least two cups of baking soda.
2. Pour at least one quart of vinegar into the large container, on top of the baking soda.
3. Observe the chemical reaction that occurs. DO NOT disturb the container. You should see a lot of bubbles form.
4. Once the reaction settles down, carefully blow bubbles into the container. Try to blow the bubbles straight across the top of the container, perhaps a foot above the container. Do not blow down into the container, because doing so can displace the carbon dioxide that is produced.
5. One or more of the bubbles should settle into the container, but float on top of the gas formed.
6. Note that the bubble tends to slowly sink over time. At the same time, the bubble increases in volume.
7. Light a candle and **CAREFULLY** lower it into the container. Note the level at which the candle goes out.
8. Use your hands to "scoop" out the gas in the container.
9. See if you can get a bubble to float in the container after the gas has been scooped out.
10. Light the candle and slowly lower it into the container. Does it go out this time?

Disposal:

The residue in the container can be rinsed down the drain with plenty of water.

Discussion:

When the baking soda and vinegar are mixed, carbon dioxide gas is formed:

$$NaHCO_3(s) + CH_3COOH(aq) \rightarrow CH_3COONa(aq) + H_2O + CO_2(g) \qquad (1)$$

Carbon dioxide (ρ = 1.8 g/L) is denser than air (ρ = 1.2 g/L) and sinks, creating an invisible layer of gas in the container. The soap bubbles, containing mostly air, float on this denser gas layer because their overall density is lower than that of carbon dioxide. Some observers may wonder why the bubble sinks, because it contains carbon dioxide from exhaled breath. However, only 4% of the gas we exhale is CO_2 so that the bubble contains mostly air.

Over time, the bubbles blown into the container slowly sink. With careful observation, it can be seen that the bubbles grow in volume as they sink. This occurs by the following mechanism. Carbon dioxide is at high concentration below the bubble, and at low concentration inside the bubble. The thin aqueous film of the bubble contains water, and $CO_2(g)$ generated below the bubble (Equation 1), enters the bubble film:

$$CO_2(g) \rightarrow CO_2(aq) \quad (2)$$

The $CO_2(aq)$ reacts with hydroxide ion in the bubble solution, which contains detergent and is basic:

$$CO_2(aq) + OH^-(aq) \rightarrow HCO_3^-(aq) \quad (3)$$

The bicarbonate ions formed in this manner diffuse through the aqueous bubble membrane to the inside surface of the bubble. At the inner bubble surface, the reverse of Equation 3 occurs where the carbon dioxide concentration inside the bubble is low:

$$HCO_3^-(aq) \rightarrow CO_2(g) + OH^-(aq) \quad (4)$$

With Equations 2 and 3 happening on the outer bubble surface and Equation 4 occurring on the inner bubble surface, this overall creates a pathway whereby CO_2 moves into the bubble. Notice that no similar chemical mechanism exists to move air molecules across the bubble film.

It is useful to use this demonstration when covering the gas laws. When doing so, the density of both air and CO_2 can be calculated. The calculation is started using the ideal gas law:

$$PV = n\mathrm{R}T \quad (5)$$

where P is pressure, V is volume, n is moles, R is the gas constant (0.0821 L atm^{-1} mol^{-1} K^{-1}), and T is temperature. To use Equation 5 to calculate the density of gases, we first note that the molar mass of a gas, M, is equal to its mass divided by moles:

$$M = \frac{mass}{n} \quad (6)$$

Substitution of Equation 6 into Equation 5 gives:

$$PV = \frac{mass}{M} RT \tag{7}$$

Rearranging, we have:

$$\frac{PM}{RT} = \frac{mass}{V} \tag{8}$$

Because the right-hand side of Equation 8 is equivalent to density:

$$D = \frac{PM}{RT} \tag{9}$$

Substitution of the appropriate molar masses for air (29 g mol^{-1}) or CO_2 (44 g mol^{-1}), and ambient conditions (P = 1 atm, T = 298 K) into Equation 9, we obtain the densities for air (1.2 g L^{-1}) and CO_2 (1.8 g L^{-1}).

REFERENCE

1. Kuntzleman, T. (2023). The floating soap bubble. *Chemical Education Xchange*. https://www.chemedx.org/blog/floating-soap-bubble

3

Imploding Golf Balls

TOPIC: volume measurement, pressure

Materials:

- Several golf balls.
- Liquid nitrogen.
- Caliper.
- Styrofoam containers or very large Styrofoam bowls. Do **NOT** use a glass Dewar!
- Cryogloves, face shield, and other personal protective equipment for use of liquid nitrogen.
- Tongs.

Hazards:

Cryogloves, face shield, and other appropriate personal protective equipment should be worn when handling liquid nitrogen. Do not immerse golf balls into a glass Dewar filled with liquid nitrogen. The golf balls snap in two with enough force to crack the Dewar, causing it to implode.

Procedure:

1. Using the caliper, find the diameter, *d*, of each golf ball.
2. Add the golf balls to the bowl or Styrofoam container and pour liquid nitrogen over the balls until they are fully immersed.
3. Keep yourself and other materials at least 3 ft away from the golf balls as they are cooling. Do not handle the golf balls with anything other than tongs after they have been immersed in the liquid nitrogen.
4. Allow the golf balls to cool for 7 to 10 minutes. Continue to replenish the nitrogen so that the golf balls remain fully immersed.
5. During the time as the golf balls are cooled, use the radius (half the diameter) of each golf ball to calculate the volume of each golf ball. Use the formula for the volume, *V*, of a sphere:

   $$V = {}^{4}/_{3}\,\pi r^3 \qquad (1)$$

   Where it is recalled that the radius of a sphere, *r*, is half the diameter ($r = d/2$).

6. After about 7 to 10 minutes, the golf balls will snap in two pieces. It is occasionally observed that some golf balls do not break in two. Use the tongs to remove the unbroken golf balls from the liquid nitrogen and place them in a cardboard box until they warm to room temperature.
7. Use tongs to remove the pieces of broken golf balls. Use the caliper to find the diameter of the golf balls that snapped in two. Make sure to do this while the golf balls are still very cold.
8. Use Equation 1 to calculate the volume of the chilled golf balls that snapped in two.

Discussion:

The composition and construction of golf balls vary.[1] Nevertheless, many have an inner core that contains either polybutadiene rubber or other polymers. Some golf balls have as many as four different layers of material. The outer layer of a golf ball generally contains harder polymer materials such as Teflon or Surlyn. The softer polymers inside the golf ball are generally rubbery at room temperature. Such materials are good at storing the energy of impact from a golf club, allowing the ball to travel far.

Rubbery polymers undergo significant changes upon cooling. At room temperature, golf balls have an elastic modulus of about 140 MPa.[2] When chilled below their glass transition temperature, or T_g, these polymers change from an elastic state to a glassy state. As the name implies, polymers become stiff and brittle in the glassy state, and the elastic modulus increases one or two orders of magnitude[3] (in this case, to as much as 14 GPa). Below T_g, polymers are subject to brittle failure, a form of spontaneous breakage.

The temperature of liquid nitrogen is −196 °C, while glass transition temperatures for polybutadiene and other rubbery polymers usually fall between −100 °C and −75 °C. Therefore, the materials inside golf balls that are immersed in liquid nitrogen are cooled to well below T_g. In this experiment, the radius of golf balls is routinely observed to decrease enough to shrink the volume of the golf balls by as much as 5%. This degree of constriction of polymer material in the glassy state is very likely the main contribution to the breaking of chilled golf balls. This argument can be made quantitative in the following way. Let's use the results of one experiment where the radius of a golf ball shrinks from 2.14 to 2.10 cm upon cooling. The strain, ε, on the golf ball is calculated as:

$$\varepsilon = \frac{|\Delta r|}{r_0}$$

where Δr is the change in radius on cooling and r_0 is the original radius. Thus, $\varepsilon = 0.0187$ for the golf ball in the experiment above. The stress, σ, on the golf ball is the product of the strain and the elastic modulus, E, for the material:

$$\sigma = \varepsilon E$$

Using 1.4 to 14 GPa as the elastic modulus for golf balls, we calculate that the stress on the golf balls due to cooling is about 26 to 260 MPa. This internal stress can be thought of as equivalent to a pressure of 250 to 2500 atmospheres (1 atm ≈ 0.101 MPa), far exceeding atmospheric pressure as a result of cooling. It's no wonder why they snap in two!

REFERENCES

1. McKay, S. E., Robbins, T., & Cole, R. S. (2008). Modern sport and chemistry: What a golf fanatic should know. *Journal of Chemical Education, 85*, 1319–1322.
2. Michal, G. M., & Novak, M. D. (2001). *Advanced Material Processing, 159*, 31–35.
3. Liska, J. W. (1944). Effect of low temperatures on Young's modulus of elastomers. *Industrial & Engineering Chemistry, 36*, 40–46.

4

Heavy Water

TOPIC: isotopes, density, molar mass, moles

Materials:

- Deuterium oxide (available for purchase several places online).[1,2]
- Ice cube tray.
- Freezer.
- Beaker filled with water.

Procedure:

1. At least one day ahead of the demonstration, fill one section of an ice cube tray with deuterium oxide (this will take roughly 25 to 40 mL).
2. Using great care not to mix the samples, fill a few other sections of the ice cube tray with deionized water (or tap water).
3. Allow the ice cubes to freeze.
4. Fill a beaker or glass container that is large and tall (to allow for clear observation of floating/sinking behavior of ice cubes) with deionized (or tap) water.

5. Add one of the regular ice cubes to the container of water.
6. Add the ice cube made of deuterium oxide to the container of water.
7. Compare the floating and sinking behavior of the two ice cubes.

Discussion:

This demonstration works well when introducing the topic of isotopes. However, it can also be used quantitatively in a manner that ties together the concepts of isotopes, density, and molar mass.

Hydrogen consists of two stable isotopes: hydrogen 1 (^{1}H or H, also known as protium) and hydrogen 2 (^{2}H or D, also known as deuterium). Protium has only a single proton in the nucleus, while deuterium has both a proton and a neutron. Thus, deuterium is about twice as heavy as protium. Almost all of the hydrogen found in various chemical compounds is in the form of protium (99.985%), while a tiny fraction (0.0015%) is deuterium. Therefore, in any given sample of water, essentially all of the water is in the form of H_2O. However, a tiny fraction of both HDO and D_2O also exist. To an excellent approximation, the water we use on a regular basis contains almost entirely H_2O. The term *regular water* will be used to refer to this type of water. However, in this demonstration a sample that is almost 100% D_2O is also used. Water in which both hydrogen atoms are deuterium is commonly referred to as *heavy water*.

Using the molar masses of protium (1.0 g mol^{-1}), deuterium (2.0 g mol^{-1}), and oxygen (16.0 g mol^{-1}), we can calculate the molar masses of regular water (18.0 g mol^{-1}) and heavy water (20.0 g mol^{-1}). In this demonstration, regular ice is observed to float, while heavy ice is observed to sink. The densities of normal ice (0.92 g cm^{-3}) and regular liquid water (1.00 g cm^{-3}) explain why regular ice floats. We can estimate the density of heavy ice by imagining that we replace each molecule of H_2O in ice with D_2O:

$$\frac{0.92\ \mathrm{g\ H_2O}}{\mathrm{mL}} \times \frac{1\ \mathrm{mol\ H_2O}}{18.0\ \mathrm{g\ H_2O}} \times \frac{1\ \mathrm{mol\ D_2O}}{1\ \mathrm{mol\ H_2O}} \times \frac{20.0\ \mathrm{g\ D_2O}}{1.0\ \mathrm{mol\ D_2O}} = 1.02\ \mathrm{g\ mL^{-1}}$$

This estimation indicates that the density of heavy ice is greater than that of regular water. Therefore, heavy ice should sink in regular water, consistent with observations.

Students often ask if it is okay to drink heavy water. While not recommended, it appears to be safe in small amounts but becomes toxic if concentrations reach about 20% in the body.[3]

REFERENCES

1. Fisher Scientific. (n.d.). Deuterium oxide, for NMR, 99.8 atom % D. https://www.fishersci.com/shop/products/deuterium-oxide-nmr-99-8-atom-d-thermo-scientific/AC166301000
2. United Nuclear. (n.d.). Deuterium oxide. https://unitednuclear.com/chemicals-metals-c-69/deuterium-oxide-p-135.html
3. Mosin, O., & Ignatov, I. (2015). Phenomenon of biological adaptation to heavy water. *Journal of Health, Medicine, and Nursing, 9*, 72–110.

5

Lithium, an Alkali Metal

TOPICS: characteristics of an alkali metal, chemical reactions, flame test

Materials:

- 3 V button battery such as a CR3032.
- Pliers.
- Acid–base indicator such as phenolphthalein.
- 250 mL Erlenmeyer flask filled with about 225 mL of water.
- Butane lighter with nozzle.

Hazards:

Although researchers have opened many lithium batteries without incident,[1] disassembly carries risk of fire or explosion and should only be done with appropriate precaution. This step should only be carried out by instructors or trained professionals, never by students. A fire extinguisher should be close by. Gloves and eye protection should be worn.

Procedure:

1. Add acid–base indicator to the beaker of water.
2. Wearing gloves and goggles, use the pliers to pry open the battery. When doing so, place the pliers right on the lip of the crimp, and press the battery against a solid surface (see https://www.youtube.com/shorts/keqdugjSk04).
3. Upon opening the battery, remove the side of the battery that contains black manganese dioxide. Also remove the plastic membrane covering the lithium side.
4. Drop the lithium side of the battery into the water. Note the formation of hydrogen gas bubbles and the color change in the acid–base indicator.
5. Using the lighter, attempt to ignite the bubbles of hydrogen gas.

Discussion:

While this demonstration can find use in a variety of contexts, I use it when discussing the various families in the periodic table. Thus, the properties of lithium seen in this demonstration are used to highlight the properties of the alkali metals. If very small samples of metallic sodium or metallic potassium are available, it is useful to demonstrate the properties of these elements as well in addition to the lithium extracted from the battery.

Lithium is an alkali metal that finds use as the anode in several battery systems. In button-cell batteries, the chemical reaction that powers the battery is:[2]

$$Li(s) + Mn^{IV}O_2(s) \rightarrow LiMn^{III}O_2(s) \tag{1}$$

where it is noted that the lithium metal is oxidized to Li^+ in the $LiMn^{III}O_2$ product. As with all alkali metals, lithium reacts with water to form base (alkali) as well as hydrogen gas:

$$2\,Li(s) + 2\,H_2O(l) \rightarrow 2\,LiOH(aq) + H_2(g) \tag{2}$$

Thus, when the lithium is added to the water, bubbles are observed, and the phenolphthalein indicator changes to a pink color.

The bubbles of hydrogen gas that are produced in this reaction can be ignited. With care, a scarlet-red flame can be sustained. The bubbles ignite due to the reaction between hydrogen gas and oxygen in the air:

$$2\ H_2(g) + O_2(g) \rightarrow 2\ H_2O(g) \tag{3}$$

The scarlet color of the flame is due to the emission of light from electron transitions in lithium atoms that are excited via the thermal energy released by this highly exothermic reaction.

REFERENCES

1. Martínez, V. A., & Ibanez, J. G. (2020). All roads lead to Rome: Triple stoichiometry with a lithium battery. *Journal of Chemical Education, 97*, 4103–4107.
2. Treptow, R. S. (2003). Lithium batteries: A practical application of chemical principles. *Journal of Chemical Education, 80*, 1015–1020.

6

Chemistry of the Statue of Liberty

TOPICS: oxidation–reduction reactions, metal complexes, transition elements with multiple oxidation numbers, general interest

Materials:

- Copper coins.
- Steel wool.
- 9 V battery.
- Aluminum foil.
- Paper towel.
- Household ammonia.
- Table salt.
- Plastic plate.
- 3% hydrogen peroxide.

Hazards:

Wear gloves and safety goggles. Household ammonia releases irritating fumes, so work in a well-ventilated area.

Procedure:

1. Buff the copper coins with steel wool until shiny.
2. Add some household ammonia to the plastic plate. Sprinkle in some salt and stir to dissolve.
3. Soak a paper towel (about 8 cm on each side) in the ammonia–salt mixture.
4. Squeeze out the excess fluid from the paper towel.
5. Place a square piece of aluminum foil about 10 cm on each side on a flat surface.
6. Fold the paper towel so that it can lie flat on the aluminum foil square.
7. Place a copper coin face down on the paper towel, ensuring that it does not touch the foil square.
8. To complete the circuit, simultaneously touch the positive terminal of the 9 V battery to the copper coin and the negative terminal to the aluminum foil. Press firmly to ensure electrical contact.
9. After pressing down hard on the penny for about 5 s, remove the penny.
10. To increase the likelihood that a patina will form on the penny, repeat steps 7 to 9 up to four times. Each time, expose the same side of the penny to a fresh portion of the paper towel.
11. Note that after each exposure, a blue-green impression forms on the paper towel.
12. After a few exposures, place the penny treated side facing up and observe. A patina should develop over the course of a few minutes. The reaction will continue for several hours or days.
13. Show the audience the blue-green impression that formed on the paper towel.
14. Add drops of 3% hydrogen peroxide to the blue-green impression on the paper towel. Note the formation of an orange-brown product.

Discussion:

While this demonstration can be used in a variety of contexts, I tend to use it to reinforce the idea that many transition elements exist in multiple oxidation states. The materials involved in this experiment are not well-suited for display to a large audience. However, the use of a document camera is helpful in this regard.

Copper displays rich chemistry. Copper metal has a characteristic red-orange metallic luster. When oxidized, copper exists as either Cu^{+} or Cu^{2+}, the former of which tends to form orange compounds, while the latter forms blue or green compounds.

Metallic copper slowly oxidizes upon contact with various components in the atmosphere such as $O_2(g)$, $CO_2(g)$, or $SO_2(g)$. For example, metallic copper reacts with oxygen to form cuprite (Cu_2O), which is red or pink in color:

$$2\ Cu(s) + \tfrac{1}{2}\ O_2(g) \rightarrow Cu_2O(s) \qquad (1)$$

Note that copper exists as copper (I) in cuprite. Further reaction of cuprite with oxygen results in tenorite (CuO), a copper (II) compound that is black in color:

$$Cu_2O(s) + \tfrac{1}{2}\ O_2(g) \rightarrow 2\ CuO(s) \qquad (2)$$

Copper also reacts with a variety of atmospheric compounds, becoming oxidized to form blue-green coatings called copper patinas. The blue-green color of a copper patina arises from the Cu^{2+} ion that is present in all such patinas. Perhaps the most famous copper patina is the green covering on the Statue of Liberty, which is made of copper. Copper patinas are generally composed of a very large number of different copper compounds. However, four or five compounds are most often found in patinas. The equations below illustrate the formation of some of these more common compounds found in copper patinas:

Formation of brochantite (green):

$$4\ Cu_2O(s) + 2\ SO_2(g) + 3\ O_2(g) + 6\ H_2O(l) \rightarrow 2\ CuSO_4 \bullet 3Cu(OH)_2(s) \qquad (3)$$

Antlerite (green):

$$6\ Cu_2O(s) + 4\ SO_2(g) + 5\ O_2(g) + 8\ H_2O(l) \rightarrow 4\ CuSO_4 \bullet 2Cu(OH)_2(s) \qquad (4)$$

Atacamite (green):

$$2\ Cu_2O(s) + 2\ Cl^{-}(aq) + O_2(g) + 4\ H_2O(l) \rightarrow 2\ Cu_2(OH)_3Cl(s) + 2\ OH^{-}(aq) \qquad (5)$$

Azurite (blue):

$$6\ Cu_2O(s) + 8\ CO_2(g) + 3\ O_2(g) + 4\ H_2O(l) \rightarrow 4\ Cu_3(CO_3)_2(OH)_2(s) \qquad (6)$$

And finally, malachite (green):

$$2\ Cu_2O(s) + 2\ CO_2(g) + 2\ H_2O(l) \rightarrow 2\ Cu_2CO_3(OH)_2(s) \qquad (7)$$

The formation of cuprite, tenorite, and copper patina compounds usually takes several years. However, in this demonstration the formation of copper patinas occurs in a much shorter period of time.

In this demonstration, chemical energy provides the energy to oxidize the copper at the positive battery terminal. When the circuit is connected, the copper metal loses electrons at the penny/towel interface:

$$Cu(s) \rightarrow Cu^{+}(aq) + e^{-} \quad (8)$$

$$Cu(s) \rightarrow Cu^{2+}(aq) + 2\ e^{-} \quad (9)$$

After battery treatment, the surface of the coin is altered—likely containing Cu^{2+} ions, and it continues to react with components in the atmosphere and in the ammonia–salt mixture to form the patina. When this experiment is conducted several times, the color of the patina that forms is variable. While the composition of the patinas formed in this experiment is not known, they are likely similar to the products shown in Equations 3 to 7.

A deep blue impression forms on the paper towel. This results from the formation of a copper(II)-ammonia complex:

$$Cu^{2+}(aq) + 4\ NH_3(aq) \rightarrow Cu(NH_3)_4^{2+}\ (aq) \quad (10)$$

Adding hydrogen peroxide to this complex produces oxygen gas and an orange-brown copper(I) product:

$$2\ Cu^{2+}(aq) + H_2O_2(aq) \rightarrow 2\ Cu^{+}(aq) + O_2(g) + 2\ H^{+}(aq) \quad (11)$$

Foaming from oxygen gas production is often observed when hydrogen peroxide is added to the blue-green impression.

REFERENCE

1. Kuntzleman, T. S., Cullen, D. M., Milam, S., & Ragan, D. (2020). *Journal of Chemical Education, 97*, 2244–2248.

7

The Oxidizing Power of Nitrates:

A Chemical Fuse Demonstration

TOPICS: polyatomic ions, chemical reactions

Materials:

- Potassium nitrate (KNO_3, 10 g).
- 20 mL of water.
- Small beaker.
- Paper towel ((about 30 cm on each side).
- Lighter.
- Desiccator.

Hazards:

Wear gloves and safety goggles. Perform the combustion step in a well-ventilated area, away from flammable materials. A fire extinguisher should be nearby.

Procedure:

1. Dissolve 10 g of KNO_3 into 20 mL of water in a small beaker.

2. Immerse the paper towel in the solution of KNO_3.
3. Remove the towel, squeeze out excess solution, and allow it to dry completely—preferably overnight in a desiccator.
4. When ready to use, remove the towel from the desiccator and confirm it is completely dry.
5. Cut the paper towel into small pieces, each approximately 5 cm × 5 cm.
6. When ready for the demonstration, place a piece of the treated paper towel on a flat, flame-resistant surface in a well-ventilated area, then ignite it using a lighter.
7. If desired, light a small piece of dry, ***untreated*** paper towel on a flame-resistant surface to observe its normal burning behavior.

Discussion:

The nitrate-treated paper towel burns with noticeably increased vigor compared to untreated paper. One way to explain this result is to note that ignition of the paper causes the KNO_3 to thermally decompose according to:

$$2\ KNO_3(s) \rightarrow 2\ KNO_2(s) + O_2(g) \quad (1)$$

The oxygen gas produced in this process provides additional oxygen, which supports rapid combustion of the paper towel. It should be noted that although oxygen is released, nitrate itself can act as an oxidizer without full decomposition. Careful observation shows that the paper towel releases white "smoke" as it burns. It is likely that much of this white smoke consists of tiny solid particles of KNO_2.

The reaction outlined in Equation 1 features both nitrate (NO_3^-) and nitrite (NO_2^-) ions. This demonstration can, therefore, be used to effectively highlight the subtle but important differences between similar polyatomic ions, which is crucial for students learning chemical nomenclature. I like to point out that NO_3^- and NO_2^- differ only by the presence of a single oxygen atom, and their names also differ by a single letter. Further, many other polyatomic ions have similar characteristics. Thus, students should spend a lot of time and take great care to recognize these differences.

REFERENCE

1. Kuntzleman, T. (2023). Burning paper towels soaked in various metal nitrates. *Chemical Education XChange.* https://www.chemedx.org/blog/burning-paper-towels-soaked-various-metal-nitrates

8

Alcohol Combustion Tests

TOPIC: combustion reactions

Materials:

- Mr. Clean Magic Eraser sponge.
- Scissors.
- Methanol, ethanol, and isopropanol.
- 3 wide-mouth containers that can hold a 1 in. cube. Each container should have a tight-fitting stopper.
- Butane lighter with nozzle.
- Two pairs of tongs.
- Three test tubes.
- 1 L beaker half filled with water.

Hazards:

Ensure all alcohol containers are tightly stoppered during ignition. Keep alcohol containers sealed while conducting flame tests. Never pour alcohol onto an active or recently extinguished flame.

Procedure:

1. Fill each container with a separate kind of alcohol: methanol, ethanol, and isopropanol.
2. Cut the Mr. Clean Magic Eraser sponge into cubes that are roughly 2 cm on each side.
3. Immerse a sponge cube into each container **and stopper tightly**.
4. Using tongs, remove the sponge immersed in the container that has methanol. While removing, squeeze to remove excess liquid.

CAUTION: RESTOPPER THE CONTAINER AND REMOVE IT FROM THE DEMONSTRATION AREA BEFORE PROCEEDING.

5. Using tongs, hold the sponge at arm's length over the 1 L beaker of water and ignite it.
6. Observe the way color is distributed throughout the flame.
7. While the flame is still burning, hold a test tube with tongs and pass it back and forth through the flame for 5 s.
8. Inspect the test tube for the presence of soot (it's helpful to hold the test tube against a white background).
9. Extinguish the flame by plunging it into the beaker of water.
10. Repeat steps 4 to 9 for the other alcohols.

Discussion:

Chemical compounds that contain carbon react with oxygen to produce carbon dioxide and water in a process known as combustion. The three alcohols tested in this demonstration undergo combustion:

Methanol: $$2\,CH_3OH(l) + 3\,O_2(g) \rightarrow 2\,CO_2(g) + 4\,H_2O(g) \quad (1)$$

Ethanol: $$C_2H_5OH(l) + 3\,O_2(g) \rightarrow 2\,CO_2(g) + 3\,H_2O(g) \quad (2)$$

Isopropanol: $$2\,C_3H_7OH(l) + 9\,O_2(g) \rightarrow 6\,CO_2(g) + 8\,H_2O(g) \quad (3)$$

If the carbon-containing compound is completely converted to carbon dioxide as outlined above, **complete combustion** occurs. Complete combustion tends to occur more readily when plenty of reactant oxygen is available. Complete combustion is usually accompanied by a blue flame, such as what is commonly observed in the flame of a gas stove or Bunsen burner. When carbon-containing compounds are not completely converted into CO_2, but rather into carbon monoxide or soot, then **incomplete combustion** is occurring. The conversion of isopropanol into carbon monoxide (Equation 4) or soot (Equation 5) during incomplete combustion may be represented as:

$$C_3H_7OH(l) + 3\ O_2(g) \rightarrow 3\ CO(g) + 4\ H_2O(g) \quad (4)$$

$$2\ C_3H_7OH(l) + 3\ O_2(g) \rightarrow 6\ C(s, \text{soot}) + 8\ H_2O(g) \quad (5)$$

Because soot glows an orange-yellow color when heated, the presence of an orange-yellow color during combustion indicates that incomplete combustion is occurring. Incomplete combustion tends to occur when oxygen is limited relative to the fuel.

In this demonstration it is commonly observed that methanol burns with a blue flame (indicating complete combustion), while isopropanol burns with a yellow flame (indicating incomplete combustion). Further evidence that isopropanol undergoes incomplete combustion comes from the fact that soot is observed to be deposited onto a test tube that is inserted into an isopropanol flame. Ethanol flames tend to show a blue color near the bottom that becomes more yellow near the top of the flame as hot gases rise and oxygen is consumed. This suggests that both combustion processes take place during ethanol combustion. The observation that the flame is blue near the bottom and yellow near the top is consistent with hot reactant gases being warmed and traveling up the flame, with oxygen being consumed as gases rise in the flame. At the bottom of the flame, plenty of oxygen is available, so ethanol undergoes complete combustion quite readily at this location. However, oxygen becomes consumed as the gas mixture travels up the flame. Thus, near the top of the flame less oxygen is available and incomplete combustion is more likely.

Finally, careful observers might notice water condensation initially forming on a test tube inserted into a methanol or ethanol flame. This is consistent with the production of water in combustion. This effect does not last long because the condensation quickly evaporates as the test tube warms up.

REFERENCE

1. Landis, A. M., Davies, M. I., Landis, L., & Thomas, N. C. (2009). Magic eraser flame tests. *Journal of Chemical Education, 86*, 577–578.

9

Steel Wool Stoichiometry

TOPIC: stoichiometry

Materials:

- Extra fine steel wool (#0000).
- Scissors.
- 400 mL glass beaker.
- Butane lighter with nozzle.
- Hair dryer.

Hazards:

Wear goggles. Notice that sparks will fly out of the beaker during this demonstration, so keep the demonstration away from combustible materials. Perform the demonstration in a well-ventilated area and keep a fire extinguisher nearby.

Procedure:

1. Find the mass of the beaker.
2. Cut the piece of steel wool in half.
3. Open up the steel wool, and line the inside of the beaker with the steel wool. The steel wool should fit snugly inside the beaker.
4. Find the mass of the steel wool + the beaker; find the mass of the steel wool by difference.
5. Assuming that the steel wool is entirely iron, calculate the theoretical yield of iron (II) oxide that should be produced if the steel wool completely reacts according to:

$$2\ Fe(s) + O_2(g) \rightarrow 2\ FeO \qquad (1)$$

6. Place the beaker in a well-ventilated area clear of combustible materials.
7. Using a lighter, ignite the steel wool.
8. **Use caution in this step:** Use the hair dryer to gently blow fresh air onto the burning steel wool. You want to blow enough air to make the steel wool glow, but not so much that sparks (or the steel wool itself) flies out of the beaker.
9. Try to make sure that all of the steel wool reacts, so ensure that all of the wool glows orange at some point during the reaction. It is also noted that the iron oxide has a slight blue tint. This blue color has been repeatedly observed under the described conditions. Therefore, ensure that all of the wool has been converted to light blue color.
10. The reaction releases a lot of heat and light, and sparks will be generated. Continue to blow air onto the wool until the reaction stops.
11. Allow the beaker and its contents to cool.
12. Find the mass of the product iron oxide plus the beaker.
13. Find the mass of the iron oxide produced by difference.
14. Calculate the percent yield of the iron oxide collected, according to

$$\%\ \text{yield} = \frac{\text{FeO collected (actual yield)}}{\text{theoretical yields of FeO}} \times 100\%$$

Disposal:

After the steel wool has cooled to room temperature it may be placed in a trashcan.

Discussion:

Though iron can form multiple oxides, experimental data from this demonstration consistently indicate formation of FeO. Therefore, this experiment likely involves the reaction between the iron in the steel wool and oxygen in the air:

$$2\ Fe(s) + O_2(g) \rightarrow 2\ FeO \qquad (1)$$

This experiment allows for a connection between several topics in stoichiometry. First, it allows students to calculate the theoretically expected amount of FeO produced. After running the reaction and weighing the product, the percent yield of FeO can also be found.

If desired, the information gathered in this experiment can be used to show that the product is very likely FeO, and not Fe_2O_3. This is done by noting that the mass of iron in the steel wool is the same as the mass of iron in the iron oxide product.

Subtracting the mass of iron originally used from the mass of iron oxide produced yields the mass of oxygen in the product. Converting these masses to moles allows calculation of the stoichiometric ratio of iron to oxygen in the compound. If care has been taken to ensure complete reaction, the ratio is often close to 1:1, indicating the formation of iron (II) oxide.

REFERENCE

1. Kuntzleman, T. (2015). Put a spark in your lesson on stoichiometry! *Chemical Education Xchange.* https://www.chemedx.org/blog/put-spark-your-stoichiometry-lesson

10

The Law of Electric Charge

TOPIC: law of electric charge

Materials:

- Several Styrofoam plates (at least six).
- Hair dryer (optional).

Procedure:

NOTE: This experiment might not work well in a humid environment. If humidity is a problem, try thoroughly drying your hair and the plates very well with a hair dryer prior to trying the experiment. Dry winter air increases the effectiveness of this demonstration.

1. Rub a Styrofoam plate on your hair back and forth several times to charge it.
2. Place the charged plate onto a flat surface.
3. Take a second plate and rub it on your hair back and forth several times to charge it.
4. Place this second charged plate on top of the first plate.

5. Continue with steps 3 and 4 to make a stack of charged plates. Because all the plates acquire the same electric charge, electrostatic repulsion will eventually make it difficult to place another plate on top of the stack.
6. Charge one more plate by rubbing it in your hair. Explain to the audience that, like the others, this plate is negatively charged, and since your hair is positively charged, the two now have opposite charges. While saying the phrase, "opposite charges attract," stick the plate to the positively charged hair on your head. For a fun effect, continue to lecture with the plate stuck to your head.

Discussion:

This is a helpful demonstration to introduce students to the forces between subatomic particles, especially protons and electrons.

Qualitatively speaking, the *law of electric charge* states that opposite electric charges attract, while like electric charges repel. The greater the magnitude of the charges, the stronger the attraction or repulsion. Additionally, the closer the charges are to one another, the stronger the force between them.

This law assures us that a positive charge will repel another positive charge, and a negative charge will repel another negative charge. On the other hand, positive and negative electric charges attract one another. Because protons and electrons have opposite electric charges, they attract one another. The flip side of this is that two electrons repel each other. Protons repel each other as well. It is natural to question why the protons in the nucleus of atoms don't repel one another, making the nucleus unstable. The answer to this is the presence of neutrons in the nucleus. In a sense, neutrons are the "glue" that hold the protons together in the nucleus (via the strong nuclear force).

In this demonstration, as the plates are rubbed through your hair, electrons travel from your hair onto the plate. The excess electrons give the plate a negative charge. Your hair, having lost electrons (but not protons), becomes positively charged. Therefore, the negative plates repel each other. However, the negatively charged plate is attracted to your positively charged hair.

11

Reaction between Sodium Bicarbonate and Acetic Acid

TOPIC: stoichiometry, percent yield

Materials:

- Electronic digital balance.
- Alka-Seltzer tablet or similar.
- 250 mL Erlenmeyer flask.
- 25 to 50 mL of vinegar.
- Watch glass.

Procedure:

1. Obtain an effervescent antacid tablet.
2. Note and record the amount of sodium bicarbonate ($NaHCO_3$) contained in each tablet as claimed by the manufacturer. Alka-Seltzer no longer lists the amount of $NaHCO_3$ on the package, but it is known to contain 1916 mg of $NaHCO_3$.[1]
3. Add 25 to 50 mL of vinegar to the Erlenmeyer flask.

4. Find and record the mass of the entire assembly: vinegar, Erlenmeyer flask, watch glass, and tablet.
5. Carefully add the effervescent antacid tablet to the vinegar in the Erlenmeyer flask.
6. To minimize loss of water as the reaction proceeds (there will be a lot of fizzing), place the watch glass over the mouth of the Erlenmeyer flask.
7. When the tablet has dissolved and the reaction has completed (there will be no more fizzing), find and record the total mass of the Erlenmeyer flask, watch glass, reacted tablet, and vinegar.
8. The loss in mass recorded between steps 4 and 7 can be assumed (to a good approximation) to be due entirely to CO_2 escape.

Discussion:

Effervescent antacid tablets (like Alka-Seltzer) generally contain $NaHCO_3$. When these tablets are dropped in water, acids within the tablet react with the $NaHCO_3$ to form carbon dioxide:

$$NaHCO_3(aq) + H^+(aq) \rightarrow Na^+(aq) + H_2O(l) + CO_2(g) \tag{1}$$

The acids contained in an Alka-Seltzer tablet do not provide a sufficient number of protons to completely consume all the $NaHCO_3$ present.[1] Therefore, the Alka-Seltzer tablet is reacted in vinegar (which contains 5% acetic acid) to provide a sufficient excess of protons.

Alka-Seltzer tablets contain 1916 mg of $NaHCO_3$. Therefore, the theoretical yield of CO_2 produced from complete reaction of $NaHCO_3$ via Equation 1 can be determined as follows:

$$1.916\ \text{g NaHCO}_3 \times \frac{1\ \text{mol NaHCO}_3}{84.01\ \text{g NaHCO}_3} \times \frac{1\ \text{mol CO}_2}{1\ \text{mol NaHCO}_3} \times \frac{44.01\ \text{g mol CO}_2}{1\ \text{mol CO}_2} = \mathbf{1.004\ g\ CO_2}$$

The mass of CO_2 removed from the tablet (the actual yield) as a result of addition of acid is found in this experiment. Therefore, the percent yield of CO_2 in this reaction can then be calculated:

$$\%yield = \frac{actual\ yeild}{theoreticd\ yeild} \tag{2}$$

REFERENCE

1. Chen, Y.-H., & Yaung, Y.-F. (2002). *Journal of Chemical Education, 79*, 848–850.

12

Hydrated Compounds

TOPIC: hydrated compounds

Materials:

- Test tube, ring stand, and clamp.
- A colored hydrated salt such as $CuSO_4 \cdot 5H_2O$ and/or $CoCl_2 \cdot 6H_2O$.
- Scoopula.
- Bunsen burner or propane torch.
- Water.

Procedure:

1. Attach the test tube to the ring stand with the clamp.
2. Use the Scoopula to place a small amount of the chosen hydrated salt into the test tube.
3. Note the color of the hydrated salt.
4. Gently heat the salt in the test tube. Observe any color change and condensation of water on the inner surface of the test tube.

5. When the test tube has cooled, carefully add a small amount of water to the test tube. Note any color changes. Alternatively, observe whether the salt changes color as it cools and reabsorbs water from the atmosphere.

Discussion:

Hydrated salts such as $CuSO_4 \cdot 5H_2O$ and $CoCl_2 \cdot 6H_2O$ have water molecules incorporated into their ionic lattices in a specific molar ratio. These waters of hydration can be driven out of the crystal by heating, forming the *anhydrous* salt:

$$CuSO_4 \cdot 5H_2O \rightarrow CuSO_4 + 5\ H_2O(g) \qquad (1)$$

$$CoCl_2 \cdot 6H_2O \rightarrow CoCl_2 + 6\ H_2O(g) \qquad (2)$$

The process is reversible, so that water can be reincorporated back into the crystal:

$$CuSO_4 + 5\ H_2O(l) \rightarrow CuSO_4 \cdot 5H_2O(s) \qquad (3)$$

$$CoCl_2 + 6\ H_2O(l) \rightarrow CoCl_2 \cdot 6H_2O(s) \qquad (4)$$

The loss or gain of water changes the chemical environment around the metal ion in the crystal. These structural changes are responsible for the change in color observed. $CuSO_4$ switches between blue (hydrated) and white (anhydrous), whereas $CoCl_2$ changes between pink (hydrated) and blue (anhydrous). The condensation observed during heating results from water vapor driven off from the salt that recondenses on the cooler part of the tube.

13

Limiting Reagents: The Reaction Between Mg and HCl

TOPIC: limiting reagent

Materials:

- 15 mL of 3 M HCl. In lieu of 3 M HCl, 100 mL of vinegar may be used. It should be noted that vinegar contains acetic acid ($HC_2H_3O_2$). The reaction proceeds more slowly in vinegar than in 3 M HCl.
- Large test tube, test tube clamp, and ring stand (for use with 3 M HCl).
- 250 mL beaker (for use with vinegar).
- 5 cm of Mg ribbon, precisely weighed. Typically, this will have a mass of approximately 0.01 g.

Procedure:

1. Use the test tube clamp to attach the test tube to the ring stand.
2. If using vinegar, skip step 3.

3. Fill the test tube with 15 mL of 3 M HCl. If students have not yet learned how to calculate molar concentrations, they will need to be informed that the solution contains 0.03 moles of HCl.
4. If using HCl, skip step 5.
5. Fill the 250 mL beaker with 100 mL of vinegar.
6. Display the piece of Mg ribbon to the students, and then drop the solid Mg into the chosen solution of acid (either HCl in the test tube or vinegar in the beaker).
7. As the reaction proceeds, explain to students that the reaction is represented by Equation 1 (if HCl is used) or Equation 2 (if vinegar is used):

$$Mg(s) + 2\ HCl(aq) \rightarrow MgCl_2(aq) + H_2(g) \quad (1)$$

$$Mg(s) + 2\ HC_2H_3O_2(aq) \rightarrow Mg(C_2H_3O_2)_2(aq) + H_2(g) \quad (2)$$

8. Ask the students to continue to observe the reaction, and to determine which reactant is limiting (Mg or the respective acid), based on their observations.

Discussion

Pragmatically, the *limiting reagent* in a chemical reaction is the reactant that is completely consumed during the reaction, while some of the *reagent in excess* is left over after the reaction is complete. The Mg ribbon is observed to completely dissolve in either source of acid (3 M HCl or the $HC_2H_3O_2$). Based on this, it can be concluded that Mg is the limiting reagent, while the acid is the reagent in excess in both cases.

It is useful to make these observations quantitative with students. We'll go through the calculations for the reaction in 3 M HCl first and follow with calculations for the reaction in vinegar. Note that if students have not yet covered solution concentration, the moles of HCl or $HC_2H_3O_2$ present in either case may simply be given to the students.

<u>In 3 M HCl:</u> First we calculate the moles of Mg: A strip of Mg with a mass of 0.01 g contains 0.0004 mol of Mg (0.01 g/24.3 g mol^{-1}). Next, we find the moles of HCl in 15 mL of 3 M HCl:

$$(3\ \text{moles HCl}\ L^{-1})(0.015\ L) = 0.045\ \text{mol HCl}$$

From Equation 1, we see that to consume 0.045 mol of HCl, 0.0225 mol of Mg is required:

$$0.045 \text{ mol HCl} \times \frac{1 \text{ mol Mg}}{2 \text{ mol HCl}} = 0.0225 \text{ mol Mg}$$

Because only 0.0004 mol of Mg was added to the test tube, it was the limiting reagent and was completely consumed in the reaction, consistent with observations.

In vinegar: The calculation for the moles of Mg is the same as before: A strip of Mg with a mass of 0.01 g contains 0.0004 mol of Mg (0.01 g/24.3 g mol^{-1}). The calculation of moles of $HC_2H_3O_2$ in 100 mL of vinegar is a bit more involved. Most bottles of vinegar contain a 5.0% solution of acid, and the density of commercial vinegar is roughly 1.01 g mL^{-1}. From these pieces of information the moles of $HC_2H_3O_2$ may be calculated:

$$\frac{5.0 \text{ g } HC_2H_3O_2}{100 \text{ g vinegar}} \times \frac{1.01 \text{ g vinegar}}{1 \text{ mL vinegar}} \times 100 \text{ mL vinegar} \times \frac{1 \text{ mol } HC_2H_3O_2}{60.1 \text{ g } HC_2H_3O_2} = 0.084 \text{ moles } HC_2H_3O_2$$

Using Equation 2, We see that to consume 0.084 mol of acid, 0.042 mol of Mg is required:

$$0.084 \text{ mol } HC_2H_3O_2 \times \frac{1 \text{ mol Mg}}{2 \text{ mol } HC_2H_3O_2} = 0.042 \text{ mol Mg}$$

Once again, we see that because only 0.0004 mol of Mg was added to the test tube, it was the limiting reagent and was completely consumed in the reaction, consistent with observations.

14

Conductivity of Solutions

TOPIC: electrolytes

Materials:

- 9 V battery with battery clip.
- Insulated copper wire.
- Black electrical tape.
- Ruler or wooden paint stirrer.
- 12V DC piezoelectric buzzer.
- Heat-shrink wire connectors.
- Hair dryer.
- 250 mL beakers rinsed well with deionized water.
- Deionized water.
- NaCl.
- Sugar.
- (Optional) Several brands of bottled water, including Aquafina.
- Tap water.

Procedure:

Part A –Making the conductivity meter:

1. Tape the 9 V battery to one end of the ruler or wooden paint stirring stick.
2. Connect the battery clip to the 9 V battery.
3. To one of the battery leads, connect some copper wire and run the wire all the way to the other end of the ruler, leaving 1 to 2 cm of the wire extending from the end of the ruler. Cover the connection between the battery lead and wire with a heat-shrink wire connector, and heat with a hair dryer to seal the connection.
4. To the other battery lead, connect some copper wire that extends roughly halfway down the ruler. Cover the connection between the wire and this second lead with the heat-shrink wire connector.
5. Connect one lead of the piezoelectric buzzer to the open end of the wire, and seal with a heat-shrink wire connector.
6. To the other lead of the piezoelectric buzzer, connect enough wire to extend to the end of the ruler, leaving 1 to 2 cm extending from the end of the ruler.
7. Seal this last connection with a heat-shrink wire connector.
8. Secure the wires and piezoelectric buzzer with electric tape. Be sure to secure the two wire leads extending from the end of the ruler so that they do not touch each other, but can test various liquids.

Part B—Testing deionized water, tap water, salt water, vinegar, and sugar water:

1. Fill three beakers half full with deionized water. Add salt to one of these beakers, and sugar to another beaker. Make no additions to the third beaker.
2. Fill the fourth beaker with tap water.
3. Insert the two leads from the conductivity meter into the deionized water, noting any sound that is produced.
4. Repeat step 3 for tap water, salt water, vinegar, and sugar water. Rinse the conductivity meter in deionized water after each test.

Part C—Testing various brands of bottled water:

1. Add various brands of water to separate beakers until each beaker is half full.
2. Display each water bottle behind the beaker that contains each respective brand of water.
3. Insert the two leads from the conductivity meter into each beaker of water, noting any sound that is produced. Make certain to rinse the leads with deionized water in between each test.

Discussion:

The buzzer on the conductivity meter constructed for these demonstrations only sounds if the solution allows electric current to flow between the two wire leads, completing the circuit. Because pure water contains very few ions to carry electric charge, the buzzer does not sound when it is immersed in highly purified water such as Aquafina (purified by reverse osmosis), distilled water, or deionized water.

When table salt (NaCl) is added to water, the crystals of NaCl dissociate into both sodium ions and chloride ions:

$$NaCl(s) \rightarrow Na^{+}(aq) + Cl^{-}(aq) \quad (1)$$

These dissolved ions allow electric charge to move freely through the liquid. As a result, when the leads of the conductivity meter are dipped into a solution of salt water, a loud buzzing noise is heard. Substances such as NaCl that dissociate into ions when dissolved in water are called **electrolytes**. If the electrolyte in question fully dissociates into ions when dissolved into water, the electrolyte is called a **strong electrolyte**.

A buzzing sound that ranges from faint to loud is usually heard when the conductivity meter is used to test tap water. This is because some compounds dissolve in water as it passes through soil and pipes. The loudness of the buzzer gives some indication of the concentration of dissolved solids in the tap water.

Some electrolytes partially dissociate into ions when they dissolve in water. If this is the case, it is referred to as a **weak electrolyte**. Acetic acid (CH_3COOH), which is found in vinegar, is an example of a weak electrolyte:

$$CH_3COOH(aq) \leftrightarrow CH_3COO^{-}(aq) + H^{+}(aq) \quad (2)$$

Let's take a closer look at vinegar to emphasize this point. Vinegar contains roughly 5% acetic acid by mass; the remaining 95% is water. All of the acetic acid found in vinegar is dissolved in the water. Only a small fraction of the dissolved acetic acid molecules dissociate—typically around 1 in every 200. When the conductivity meter is used to test vinegar, a buzzing noise is heard, but it is not quite as loud as what is heard in the salt water. This difference in loudness can be used to emphasize the difference between strong and weak electrolytes.

Finally, some substances like sucrose (table sugar, $C_{12}H_{22}O_{11}$) do not dissociate when dissolved in water. Such substances are called **non-electrolytes**:

$$C_{12}H_{22}O_{11}(s) \rightarrow C_{12}H_{22}O_{11}(aq) \qquad (3)$$

Although dissolved sucrose molecules move freely in water, they carry no net charge. Thus, the conductivity meter does not beep when immersed in a solution of sucrose.

The conductivity meter can be used to indicate the presence of dissolved ions in various brands of water. Any brand of water that has added salts will cause the meter to sound. Common additives include $MgCl_2$ and $NaHCO_3$, but there are many others. Brands of water that contain no additives and have been distilled, deionized, or purified by reverse osmosis (RO) will not cause the buzzer to sound. Be sure to read the label: Some manufacturers purify water and then add electrolytes back in for taste. Brands of water that have been purified but contain additives after purification will cause the buzzer to sound.

15

Seashell Chemistry, Part 1

TOPICS: solubility of salts; complete and net ionic equations

Materials:

- Deionized water.
- 2 × 500 mL Erlenmeyer flasks.
- 1 × 600 mL beaker.
- Arm & Hammer Super Washing Soda (Na_2CO_3).
- Driveway Heat ($CaCl_2$).

Hazards:

There are no significant hazards associated with this experiment.

Procedure:

1. Fill each Erlenmeyer flask to about halfway with deionized water.
2. To the first flask, add 10 g of $CaCl_2$ and swirl the mixture. Note the cloudy appearance at first—this indicates that some undissolved solid remains. With continued swirling, the solid dissolves and the solution becomes transparent. Emphasize to students that the transparent appearance means the solid has fully dissolved
3. To the second flask, add 10 g of Na_2CO_3. As before, note the cloudy appearance when the solid is first added. With continued swirling, the solution becomes transparent as the solid dissolves.
4. Once both solids have dissolved, pour the contents of each flask into the 600 mL beaker. Note the formation of a cloudy-white precipitate. Students should be able to infer that a new solid has formed that does not dissolve in water.

Disposal:

The solid calcium carbonate may be collected by filtration and discarded in the trash. The remaining filtrate, consisting primarily of a dilute aqueous solution of sodium chloride, may be safely rinsed down the drain with plenty of water. Always follow local regulations and institutional guidelines for chemical disposal.

Discussion:

The main component of Driveway Heat is $CaCl_2$, while that of Arm & Hammer Super Washing Soda is Na_2CO_3. Both these solids are very soluble in water, as is demonstrated in steps 1 to 3 of the procedure outlined above. During the dissolution process, it is useful to point out that each solid consists of ions, which fully dissociate as the solid dissolves:

$$CaCl_2(s) \rightarrow Ca^{2+}(aq) + 2\ Cl^-(aq) \qquad (1)$$

$$Na_2CO_3(s) \rightarrow 2\ Na^+(aq) + CO_3^{2-}(aq) \qquad (2)$$

During the dissolution process, each mixture transitions from cloudy to transparent. The initial cloudiness occurs because undissolved solid particles are large enough to scatter light. As the solids dissolve, they separate into individual ions too small to be seen, allowing light to pass through the solution clearly.

When the two solutions are combined, a cloudy-white precipitate forms, indicating the formation of a solid. This solid is calcium carbonate, the primary chemical component of seashells:

$$Ca^{2+}(aq) + 2\ Cl^{-}(aq) + 2\ Na^{+}(aq) + CO_3^{2-}(aq) \rightarrow CaCO_3(s) + 2\ Na^{+}(aq) + 2\ Cl^{-}(aq) \quad (3)$$

Equation 3 is the **complete ionic equation** for this reaction. By focusing only on the ions involved in forming the solid precipitate, the reaction can be simplified to the **net ionic equation**:

$$Ca^{2+}(aq) + CO_3^{2-}(aq) \rightarrow CaCO_3(s) \quad (4)$$

16

Rainbow Juices

TOPICS: acids, bases, pH, indicators

Materials:

- Milk of magnesia.
- Grape juice and/or juice containing turmeric.
- Large beakers.

Hazards:

There are no significant hazards associated with this experiment.

Procedure:

1. Add 50 to 100 mL of milk of magnesia to one of the beakers.

2. Pour in some grape juice and mix well.
3. Observe the color changes that occur over a few minutes. The color will initially appear red, but will fade to violet, then blue, then gray and finally green.
4. Once the mixture turns green, add more grape juice. The color shift will repeat.
5. Repeat the experiment using juice that contains turmeric (e.g., turmeric-infused juice or turmeric solution). In this case, color changes from yellow to red will be observed.

Disposal:

Place solutions in the container indicated for aqueous base waste.

Discussion:

The violet-red color of grape juice is due to the presence of a class of pH-sensitive compounds known as anthocyanins. These compounds tend to impart red, violet, and blue color to plants, fruits, and flowers. While the color of anthocyanins can be affected by the presence of metal ions, they also change with pH. Thus, anthocyanins are acid–base indicators. Generally speaking, the color of anthocyanins is red at low pH, violet at intermediate pH, blue at slightly basic pH, and green at higher pH. Similarly, turmeric spice contains curcumin, which is also an acid–base indicator. Curcumin displays a yellow color at lower pH, orange at intermediate pH, and red at higher pH. The color of curcumin can also be changed by the presence of metal ions.

Milk of magnesia is a suspension of solid $Mg(OH)_2$ in water. When mixed with juices that contain anthocyanins, a variety of color changes occur. Grape juice is somewhat acidic and therefore has a reddish color. Upon addition of milk of magnesia to grape juice, the color slowly shifts to violet, then blue, and finally green. This color change occurs because magnesium hydroxide reacts with the acid protons in the juice, neutralizing them:

$$Mg(OH)_2(s) + 2\ H^+(aq) \rightarrow Mg^{2+}(aq) + 2\ H_2O(l) \qquad (1)$$

Thus, when milk of magnesia is added to grape juice, the color shifts from red (acidic) to violet (neutral pH). But why does the color further shift from violet to blue (slightly basic) and ultimately green (basic)?

Well, the main ingredient in milk of magnesia is magnesium hydroxide. Milk of magnesia has a cloudy-white appearance because magnesium hydroxide is a sparingly soluble salt:

$$Mg(OH)_2(s) \leftrightarrow Mg^{2+}(aq) + 2\ OH^-(aq) \qquad (2)$$

Equation 2 assures us that some hydroxide ions dissolve into the juice if any solid magnesium hydroxide is suspended in the juice. After neutralization, undissolved magnesium hydroxide continues to dissolve slowly, releasing hydroxide ions which raise the pH further. This causes the color shift to blue and finally green.

If more juice is added after the mixture turns green, the color quickly shifts back to red. This is because the protons from the fresh juice immediately consume the slight amount of hydroxide present from the little bit of magnesium hydroxide that has dissolved (Equation 2). However, over time any solid magnesium hydroxide that remains will dissolve, consuming the protons. Further dissolution of the magnesium hydroxide causes the mixture to fade to green once again.

The cycle of red to violet to blue to green can continue again upon addition of more juice—so long as there is solid magnesium hydroxide remaining in the beaker. As this solid reacts and is consumed, the mixture will fade from a cloudy to a transparent appearance.

The same processes occur in juices that contain turmeric. However, in this case the color transitions from yellow upon addition of the juice to orange as the acid is neutralized, and finally to red as more magnesium hydroxide dissolves. Once again, the cycle may be repeated upon addition of more juice that contains turmeric as long as undissolved magnesium hydroxide remains.

17

Rainbow Bottle

TOPICS: acids, bases, neutralization, density, intermolecular forces

Materials:

- Bottle capable of holding 500 mL of fluid with a tight-fitting stopper.
- 240 mL of water.
- 280 mL of 91% isopropyl alcohol.
- 50 g of NaCl.
- A universal indicator.
- 10% to 20% by weight HCl solution in a dropper bottle.
- 10% to 20% by weight NaOH solution in a dropper bottle.

Hazards:

Concentrated solutions of HCl and NaOH can cause burns. Wear gloves and goggles. Isopropyl alcohol is flammable. Keep from flame.

Procedure:

1. Pour the 91% alcohol and water into the bottle. Stopper and mix well.
2. Add the 50 g NaCl and mix until dissolved. Note that the fluid separates into two layers.
3. Place the bottle on a flat surface and add 2 to 3 mL of the universal indicator.
4. Shake the solution, noting the color of each resulting layer.
5. Open the bottle and add a one to two drops of 20% NaOH solution to the bottle. Stopper and mix well, noting the color of each layer.
6. Open the bottle and add four to five drops of 20% HCl solution to the bottle. Stopper and mix well, noting the color of each layer.
7. If desired, continue adding acid and base solutions to create a variety of colored effects.

Disposal:

The bottle may be saved for future demonstrations.

Discussion:

Due to dipole–dipole and hydrogen bonding interactions, isopropyl alcohol and water mix quite well to form a homogeneous solution. When sodium chloride is added to this solution, very strong ion–dipole forces are set up between water molecules and the sodium and chloride ions. These strong ion–dipole forces outcompete the intermolecular forces between the molecules of water and isopropyl alcohol. As a result, the alcohol molecules become "squeezed out" and form a distinct layer. This results in the mixture separating into two layers: a less dense, alcohol rich layer on top and a more dense, aqueous salt layer on bottom. Generally speaking, the universal indicator contains four different indicator dyes that, when combined, allow for the full rainbow of colors to be observed at various pH levels (Figure 1).

FIGURE 1: Color of the universal indicator at various pH levels.

In almost all cases, all these indicators preferentially dissolve in the alcohol layer instead of the water layer. Because of this, during these experiments the alcohol layer is almost always colored, but the water layer is either faintly colored or colorless. However, there is one exception: At higher pH, phenolphthalein dissolves better in water than in alcohol. Because of this, and because phenolphthalein is pink at pH 9, the bottom layer appears pink at higher pH.

REFERENCE

1. Kuntzleman, T. S., Matti, A., & Tomco, D. (2024). *Journal of Chemical Education*, *101*, 4122–4124.

18

Bouncing Batteries

TOPICS: redox reactions, cell potentials, anode, cathode, general interest

Materials:

- Fresh AA battery.
- Spent AA battery.
- Procedure:
- Hold the spent battery vertically and drop it onto a hard surface, allowing it to bounce.
- Repeat the test a few times, noting how high the battery bounces each time.
- Repeat steps 1 and 2 with the fresh battery.

Discussion:

This experiment demonstrates that used alkaline batteries bounce higher than fresh ones. This effect can be traced to the overall chemical reaction that occurs in alkaline batteries:

$$2\ MnO_2(s) + Zn(s) + H_2O(l) \rightarrow 2\ MnO(OH) + ZnO(s) \qquad (1)$$

The reaction in Equation 1 powers the battery and occurs gradually as the battery is used. As a result, a fresh battery contains mostly MnO_2, Zn, and H_2O, while a spent battery contains mostly MnO(OH) and ZnO. ZnO happens to be a somewhat elastic solid—it has even been proposed for use in golf balls to increase travel distance. The accumulation of ZnO during battery discharge, therefore, contributes to the increased bounciness of dead batteries.

Because Equation 1 represents an oxidation–reduction reaction, this demonstration provides a natural link to topics in electrochemistry. During the reaction, electrons spontaneously transfer from Zn to MnO_2, converting Zn to ZnO and MnO_2 to MnO(OH). This electron flow powers electrical devices. To prevent the battery from discharging when not in use, the Zn(*s*) and MnO_2(*s*) are kept in separate compartments. The two reactants are separated by a membrane, with the Zn sealed in the inner portion, and the MnO_2 surrounding the inner core of Zn (Figure 1).

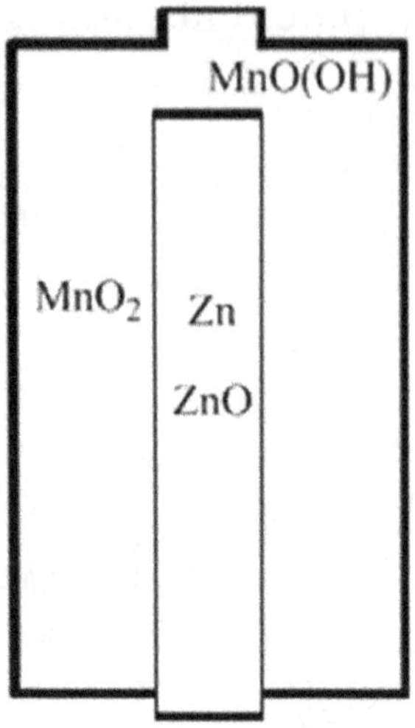

FIGURE 1: A cutaway view of an alkaline battery. The inner (anodic) portion contains Zn and ZnO; the outer (cathodic) portion contains MnO_2 and MnO(OH). The anode and cathode are separated by a membrane. When the battery is part of a completed circuit, electrons flow from the anode portion to the cathode portion.

The compartment containing Zn and ZnO is called the anode, and this is where oxidation occurs. The compartment with MnO_2 and MnO(OH) is called the cathode, and this is where reduction occurs. When the battery is not part of a circuit, no reaction occurs. However, when the battery is inserted into a powered electrical device, it completes the circuit and electrons spontaneously flow from the anode to the cathode. The half-reactions are:

$$Zn(s) + 2\ OH^-(aq) \rightarrow ZnO(s) + H_2O(l) + 2\ e^- \qquad (2), \text{anode}$$

$$MnO_2\ (s) + 2\ H_2O(l) + 2\ e^- \rightarrow MnO(OH) + 2\ OH^-(aq) \qquad (3), \text{cathode}$$

The standard reduction potential of ZnO to Zn (reverse of Equation 2) is −1.26 V, while that of MnO_2 to MnO(OH) (Equation 3) is +0.21 V. Thus, the overall standard cell potential for Equation 1 is +1.47 V, very close to the advertised 1.5 V for AA alkaline batteries.

Equation 1 represents the sum of Equations 2 and 3. While these equations describe the overall redox behavior well, a variety of manganese oxides have been reported to form in the cathode over time, and the exact chemistry remains somewhat unclear.

REFERENCE

1. Hall, J. M., Amend, J. R., & Kuntzleman, T. S. (2016). *Journal of Chemical Education, 93*, 676–680.

19

Exercise and CO_2

TOPICS: acid–base neutralization, stoichiometry, cellular respiration

Materials:

- 400 mL of 0.002 M NaOH.
- Deionized water.
- 2 × 500 Erlenmeyer flasks—these should be well rinsed with deionized water prior to use.
- Phenolphthalein indicator.
- Straws.
- Timer.

Procedure:

1. Add 200 mL of 0.002 M NaOH to both Erlenmeyer flasks.
2. Add a few drops of phenolphthalein indicator until a pink color appears.
3. Start the timer and begin exhaling steadily through a straw into one of the flasks, keeping the straw submerged.

4. Continue exhaling through the straw until the pink color disappears. When this occurs, stop the timer.
5. Engage in moderate exercise (e.g., running in place or around the room) for approximately one minute to increase your respiration rate.
6. Repeat steps 3 and 4 with the second Erlenmeyer flask.

Optional variation: Instead of NaOH and phenolphthalein, use about 1 L of Essentia water (or another alkaline water) and the universal indicator to demonstrate pH changes.

Discussion:

Humans produce CO_2 through cellular respiration according to the equation:

$$C_6H_{12}O_6 + 6\ O_2(g) \rightarrow 6\ CO_2(g) + 6\ H_2O(g) \quad (1)$$

Because CO_2 is toxic at high concentrations in the body, it is expelled through exhalation. When $CO_2(g)$ enters water, it causes an increase in proton concentration:

$$CO_2(aq) + H_2O(l) \leftrightarrow H^+(aq) + HCO_3^-(aq) \quad (2)$$

In this demonstration, the CO_2 produced by respiration neutralizes a solution of hydroxide ions:

$$H^+(aq) + OH^-(aq) \rightarrow H_2O(l) \quad (3)$$

Phenolphthalein, an acid–base indicator that is pink in basic solutions and colorless in neutral or acidic solutions, colors the hydroxide solution pink. When CO_2 is exhaled into the solution, the released protons react with hydroxide ions (Equation 3), causing the pink color to fade as the solution becomes less basic.

While it is tempting to use this demonstration to estimate the rate of human respiration, such calculations are generally inaccurate. For example, using the time required to neutralize the base (typically about 60 s), the stoichiometry of the reactions, and the volume and concentration of hydroxide, one can estimate a metabolic rate of approximately 3.6 Calories per hour based on the amount of CO_2 that reacts with the base. However, this is a significant underestimate:

$$\frac{0.1\ \text{L of OH}^-\ \text{sol'n}}{1} \times \frac{0.005\ \text{mol OH}^-}{1\ \text{L of OH}^-\ \text{sol'n}} \times \frac{1\ \text{mol H}^+}{1\ \text{mol OH}^-} \times \frac{1\ \text{mol CO}_2}{1\ \text{mol H}^+} \times \frac{1\ \text{mol C}_6\text{H}_{12}\text{O}_6}{6\ \text{mol CO}_2} \times \frac{180\ \text{g C}_6\text{H}_{12}\text{O}_6}{1\ \text{mol C}_6\text{H}_{12}\text{O}_6} \times \frac{4\ \text{Calories}}{1\ \text{g C}_6\text{H}_{12}\text{O}_6} \times \frac{1}{60\ \text{s}} \times \frac{3600\text{s}}{\text{h}} = 3.6\ \text{Cal/h}$$

This value is much lower than the typical human resting metabolic rate of about 72 Calories per hour. The discrepancy arises because much of the CO_2 exhaled escapes into the air instead of dissolving in the solution, and the reaction between CO_2 and hydroxide ions is slow.

20

Preparing Solutions

TOPIC: preparation of solutions

Materials:

- 250 mL volumetric flask.
- 100 mL volumetric flask.
- 10 mL pipette with pipette bulb.
- 6.25 g of $CuSO_4 \bullet 5H_2O$.
- Funnel wide enough to deliver solid into the 250 mL volumetric flask.
- Water.

Procedure:

1. Using the funnel, add the 6.25 g of $CuSO_4 \bullet 5H_2O$ to the volumetric flask.
2. Add water until the flask is about half full, then swirl gently to dissolve the solid.
3. Carefully add water to the calibration mark. Use a dropper to top off, if needed.
4. Stopper the flask and mix thoroughly.

5. Guide students through the calculation of the molar concentration of **$CuSO_4$** in this solution.
6. Use the pipette to transfer 10.00 mL of the solution to the 100 mL volumetric flask.
7. Fill the 100 mL flask to the mark with water. Use a dropper for precision, if necessary.
8. Stopper the flask and mix thoroughly.
9. Guide students through the calculation of the molar concentration of **$CuSO_4$** in the diluted solution.

Discussion:

This simple demonstration provides students with a visual, concrete example of preparing a solution of known concentration. In this regard, it can be helpful for students as they carry out solution concentration problems and calculations. In this demonstration, solutions of known molar concentration are prepared. The first solution is prepared by adding 6.25 g of solid $CuSO_4 \cdot 5H_2O$ (FW = 249.7 g mol^{-1}) into a 250 mL volumetric flask. The concentration of this solution is, therefore:

$$6.25\ \mathrm{g\ CuSO_4 \cdot 5H_2O} \times \frac{1\ \mathrm{mol\ CuSO_4 \cdot 5H_2O}}{249.7\ \mathrm{g\ CuSO_4 \cdot 5H_2O}} \times \frac{1\ \mathrm{mol\ CuSO_4}}{1\ \mathrm{mol\ CuSO_4 \cdot 5H_2O}} \times \frac{1}{0.2500\ \mathrm{L}}$$

$$= 0.100\ \mathrm{M\ CuSO_4}$$

The second solution is prepared by diluting 10 mL of the 0.100 M $CuSO_4$ solution to 100 mL:

$$0.100\ \mathrm{M\ CuSO_4} \times \frac{10\ \mathrm{mL}}{100\ \mathrm{mL}} = 0.0100\ \mathrm{M\ CuSO_4}$$

21

Qualitative Titration of Milk of Magnesia with HCl

TOPICS: acid–base titration, solubility, K_{sp}

Materials:

- Milk of magnesia.
- 1.0 M HCl.[1]
- 250 mL Erlenmeyer flask.
- A universal indicator.
- Water.
- Droppers.
- Digital balance.

Procedure:

1. Add about 1 g of milk of magnesia into the Erlenmeyer flask.
2. Add 100 mL of water to the Erlenmeyer flask. Mix well.
3. Add about one dropperful of the universal indicator to the flask. Swirl, noting and discussing any color changes that occur.

4. Add one to five drops (more when far from the end point and less when close to the end point) of 1.0 M HCl (or vinegar) to the flask, and swirl. Note and discuss any color changes and changes in transparency that occur within the flask.
5. Continue step 4 until the solution becomes transparent and the color change (to red, orange, or yellow) on addition of acid persists for 1 minute. When this occurs, the end point of the titration has been reached. If it is observed that the color in the flask begins to fade back to colors consistent with a higher pH, the end point has not been achieved and more acid should be added.

Discussion:

Milk of magnesia is a common antacid. Its active ingredient is magnesium hydroxide ($Mg(OH)_2$, FW = 58.3 g mol^{-1}). Magnesium hydroxide is a base that reacts with acids and neutralizes them. For example, magnesium hydroxide reacts with hydrochloric acid (HCl, FW = 36.5 g mol^{-1}) according to the equation:

$$Mg(OH)_2(s) + 2\,HCl(aq) \rightarrow MgCl_2(aq) + 2\,H_2O(l) \qquad (1)$$

Titrating $Mg(OH)_2$ with HCl reveals interesting behavior due to its low solubility in water. Magnesium hydroxide is a white solid that is mostly insoluble in water. Therefore, milk of magnesia takes on a milky-white appearance when mixed in water. However, a small portion of the solid $Mg(OH)_2$ does dissolve, releasing hydroxide ions into solution.

$$Mg(OH)_2(s) \rightarrow Mg^{2+}(aq) + 2\,OH^-(aq) \qquad (2)$$

Thus, when the universal indicator is added to a mixture of milk of magnesia in water, a blue or purple color appears (see Figure 1 on p. 59).

When acid is added to milk of magnesia mixed into water, the added H^+ ions cause a shift to a red, orange, or yellow color. This is because the added H^+ ions first react rapidly with the dissolved OH^- ions, leaving an excess of H^+ ions.

$$H^+(aq) + OH^-(aq) \rightarrow H_2O(l) \qquad (3)$$

The remaining H^+ ions go on to react slowly with the undissolved $Mg(OH)_2(s)$, neutralizing it, and causing a color shift to a more neutral color (green):

$$Mg(OH)_2(s) + 2\ H^+(aq) \rightarrow Mg^{2+}(aq) + H_2O(l) \tag{4}$$

Once the acid has been neutralized, small amounts of hydroxide dissolve from any remaining solid $Mg(OH)_2$ (Equation 2). This causes a further shift of color to aqua, blue, or violet.

Overall, the sequence of events following addition of acid to a mixture of milk of magnesia and the universal indicator is characterized by a color change from blue/aqua to red/orange/yellow and then a slow fading back to green/blue.

The equivalence point of this titration is when just enough acid has been added to completely react with all of the solid $Mg(OH)_2$, and there are many visible changes that accompany the addition of acid near the equivalence point. First, the end point is characterized by a change in color to yellow that does not fade back to blue or green, because there is no more hydroxide to consume added acid. Thus, the end point is indicated when the last drop of acid causes the mixture to gain a persistent yellow color. Second, because the acid converts the undissolved $Mg(OH)_2$ to dissolved products, the mixture becomes increasingly transparent as more and more acid is added. Thus, the reaction mixture should be completely transparent at the end point.

Note

1. Vinegar (which is a 0.84 M solution of acetic acid, $HC_2H_3O_2$) may be used as a substitute for 1.0 M HCl. Like HCl, $HC_2H_3O_2$ supplies one proton per molecule of acetic acid:

$$Mg(OH)_2(s) + 2\ HC_2H_3O_2(aq) \rightarrow Mg(C_2H_3O_2)_2(aq) + 2\ H_2O(l) \tag{5}$$

REFERENCE

1. Kuntzleman, T. S., Corts, & Schmidt. (2021). *Journal of Chemical Education, 98*, 2292–2295.

22

Club Soda Rainbow

TOPICS: solubility of CO_2 in water, pH, Le Chatelier's principle

Materials:

- Two 1 L bottles of club soda.
- A universal indicator.
- Disposable plastic pipettes.
- Two heating units.
- Beaker.
- 5 × 600 mL beakers rinsed very well with deionized water.
- Deionized water.
- (optional) Mentos candies and 2 L beaker.

Hazards:

This demonstration is generally safe when performed with appropriate caution. However, care must be taken when using hot plates and boiling liquids, as these can cause burns. Heat-resistant gloves or tongs should be used when handling hot beakers, and students should keep their hands and faces away from rising steam. Glassware may crack or shatter if it is heated unevenly or placed on a cold surface while still hot, so borosilicate glass should be used and allowed to cool gradually.

Procedure:

1. Turn the hot plates on high.
2. Open the bottles of club soda.
3. Use the plastic pipette to deliver the universal indicator to each bottle of soda. Note the color of the indicator in the club soda.
4. Reseal one of the bottles of club soda. This will be the control.
5. Use the contents of the other bottle to fill two beakers half full with club soda.
6. Fill two beakers half full with deionized water. Add the universal indicator. If the solutions do not appear yellow, thoroughly reclean the beakers with deionized water and repeat the step.
7. Place one beaker with club soda (now colored red) and one beaker with deionized water (now colored yellow) on separate hot plates. Continue to heat until boiling, and then continue heating for several minutes after boiling begins. Note any color changes that occur.
8. Pour the club soda (that is not being heated) back and forth between two beakers several times. Note any color changes.
9. Add Mentos candies to the opened club soda (the one that is NOT the control). Note any color change.
10. At the end of this demonstration, there should be several solutions that display several colors: red, orange, yellow, green, and blue. Arrange these solutions in order of the colors observed in the rainbow (ROYGB).

Discussion:

Sodas are sealed under high pressure of CO_2, which drives large amounts of CO_2 to dissolve in the beverage:

$$CO_2(g) \rightarrow CO_2(aq) \quad (1)$$

Dissolved CO_2 reacts with water to form protons and bicarbonate ion. The former lowers the pH:

$$CO_2(aq) + H_2O(l) \leftrightarrow H^+(aq) + HCO_3^-(aq) \quad (2)$$

When a soda bottle is unsealed, the pressure of CO_2 above the liquid drops significantly, causing CO_2 to escape both from the surface and via bubbles:

$$CO_2(aq) \rightarrow CO_2(g) \quad (3)$$

CO_2 loss occurs gradually under normal conditions. As CO_2 escapes, the concentration of dissolved CO_2 decreases, shifting the equilibrium in Equation 2 to the left. This reduces the H^+ concentration, raising the pH.

This demonstration shows how pH changes due to CO_2 loss can be induced by different methods and visualized using the universal indicator (see Figure 1, p. 59).

Heating causes dissolved CO_2 to escape because CO_2 solubility decreases with temperature. This CO_2 loss shifts Equation 2 left, resulting in a pH increase. This is observed as a color change from yellow to green when deionized water with indicator is heated.

The universal indicator added to freshly opened club soda typically turns red, indicating a pH below 3 due to the high CO_2 concentration. When soda is agitated by pouring or by adding Mentos, CO_2 rapidly escapes (Equation 3), and the resulting pH increase causes the color to shift from red toward orange.

Because agitation or Mentos addition only removes some CO_2, the solution remains weakly acidic. However, boiling club soda removes nearly all CO_2. Club soda (unlike most other sodas) contains basic salts, so when CO_2 is fully purged, the solution becomes basic. As a result, boiling causes color changes in sequence: red → orange → yellow → green → blue.

REFERENCE

1. Kuntzleman, T. (2020). Color changing coke and mentos. *Chemical Education Xchange.* https://www.chemedx.org/blog/color-changing-coke-and-mentos

23

Reaction Between Copper Ion and Aluminum

TOPICS: redox reactions, cell potentials, half-reactions, flame tests

Materials:

- 1.5 M copper (II) sulfate solution ($CuSO_4$).
- Aluminum foil.
- 250 mL Erlenmeyer flask.
- Table salt.
- Butane lighter with nozzle.

Procedure:

1. Add 100 mL of 1.5 M $CuSO_4$ into the Erlenmeyer flask.
2. Add some table salt to the solution of $CuSO_4$. Swirl to dissolve.
3. Add in some pieces of aluminum foil.
4. If sufficient bubbling occurs, use the lighter in an attempt to ignite these bubbles.

Discussion:

During the reaction that commences, the aluminum foil dissolves, forming a red-orange precipitate. Vigorous bubbling and the generation of a lot of heat is also observed. These observations occur because of the following chemical reaction between copper (II) ions and metallic aluminum:

$$3\ Cu^{2+}(aq) + 2\ Al(s) \rightarrow 3\ Cu(s) + 2\ Al^{3+}(aq) \qquad (1)$$

Interestingly the above reaction does not happen in the absence of chloride ions. This is because aluminum foil has an oxide coating that prevents Cu^{2+} ions from reacting with the metallic aluminum below. However, in the presence of chloride ion, this oxide coating disintegrates, allowing the reaction to occur.

The process outlined in Equation 1 is an oxidation–reduction (redox) reaction. Therefore, Equation 1 can be written as the sum of two half-reactions. The pertinent reduction half-reactions are:

$$Cu^{2+}(aq) + 2\ e^- \rightarrow Cu(s) \qquad E^0 = +0.34\ V \qquad (2)$$

$$Al^{3+}(aq) + 3\ e^- \rightarrow Al(s) \qquad E^0 = -1.66\ V \qquad (3)$$

It is noted that three times Equation 2 plus twice the reverse of Equation 3 gives the overall chemical reaction (Equation 1):

$$3\ Cu^{2+}(aq) + 6\ e^- \rightarrow 3\ Cu(s)$$

$$\underline{+\ 2\ Al(s) \rightarrow 2\ Al^{3+}(aq) + 6\ e^-}$$

$$3\ Cu^{2+}(aq) + 2\ Al(s) \rightarrow 3\ Cu(s) + 2\ Al^{3+}(aq)$$

Therefore, the cell potential for Equation 1 is +2.00 V, where the positive sign indicates the reaction is spontaneous. We also note that copper ion is reduced, while aluminum metal is oxidized in this reaction. (Also, copper is the oxidizing agent, while aluminum metal is the reducing agent).

The bubbling observed in this reaction results from the reaction between the aluminum metal and water:

$$2\ Al(s) + 6\ H^+ \rightarrow 2\ Al^{3+}(aq) + 3\ H_2(g) \quad (4)$$

This is also a redox reaction, with the pertinent reduction half-reactions:

$$2\ H^+(aq) + 2\ e^- \rightarrow H_2(g) \qquad E^0 = 0\ V \quad (5)$$

$$Al^{3+}(aq) + 3\ e^- \rightarrow Al(s) \qquad E^0 = -1.66\ V \quad (6)$$

The overall reaction (Equation 4) is obtained by adding three times Equation 5 with twice the reverse of Equation 6. Thus, the cell potential for Equation 4 is +1.66 V. In this reaction, aluminum metal is oxidized (it's the reducing agent), while protons are reduced (they are the oxidizing agent).

Finally, because the gas bubbles produced contain hydrogen gas, the bubbles can be ignited in the presence of air due to the reaction between hydrogen and oxygen to form water:

$$2\ H_2(g) + O_2(g) \rightarrow 2\ H_2O(g) \quad (7)$$

It is often the case that the flame generated has a bright green color due to the presence of electrons in copper ions being thermally excited in the resulting flame. This third reaction (Equation 7), which occurs when hydrogen gas ignites, is also a redox reaction. The half-reactions giving rise to this process are:

$$2\ H_2O(aq) + 2\ e^- \rightarrow H_2(g) + 2\ OH^-(aq) \qquad E^0 = -0.83\ V \quad (8)$$

$$4\ H^+(aq) + 4\ e^- + O_2(g) \rightarrow 2\ H_2O(g) \qquad E^0 = +1.23\ V \quad (9)$$

In this case, twice the reverse of Equation 8 plus Equation 9 yields Equation 7. The cell potential of the reaction in Equation 7 is +2.06 V.

24

Pressure in a Bottle of Soda Pop

TOPICS: ideal gas law

Materials:

- 355 mL bottles of Diet Coke or other carbonated beverage.
- Non-contact infrared thermometer.
- Balance with a capacity of at least 400 g that is capable of measuring to the nearest 0.01 g.
- Water.

Hazards:

There are no hazards associated with this experiment.

Procedure:

1. Find the temperature of the beverage.
2. Find the mass of the unopened bottle.
3. Slightly loosen the cap just enough to release the gas trapped in the headspace. Listen for the initial short, sharp hiss, then quickly tighten the cap again to prevent CO_2 loss from the liquid.
4. Find the mass of the opened bottle.
5. Calculate the mass lost from the bottle. Assume the mass lost is due entirely to the escape of CO_2 from the beverage.
6. Reopen the bottle and zero out the balance.
7. Using a dropper, fill the bottle to the brim with water. Find the mass of water required to fill the headspace. Note that this mass of water in grams is very close to the volume of the headspace in milliliters.

Disposal:

The beverage may be rinsed down the drain.

Discussion:

Bottles of soda pop are sealed under high pressures of CO_2. This is done so that large amounts of CO_2 dissolve into the beverage through the following process:

$$CO_2(g) \rightarrow CO_2(aq) \quad (1)$$

The data collected in this experiment, along with the ideal gas law (Equation 2) can be used to estimate the pressure inside the soda bottle:

$$PV = n\mathrm{R}T \quad (2)$$

In the above equation P represents pressure, V volume, n moles, T absolute temperature, and $R = 0.0821$ L atm mol^{-1} K^{-1}. Dividing both sides of Equation 2 by V yields:

$$P = \frac{nRT}{V} \tag{3}$$

Let's see how to carry out a calculation using sample data. Assume that when quickly opened and shut, 0.22 g of CO_2 escapes from the headspace of a soda bottle at 25 °C. Further assume that it is found that 32.36 mL of water was required to fill the bottle headspace.

The moles of CO_2 that escaped is found to be:

$$0.22 \text{ g } CO_2 \times \frac{1 \text{ mol } CO_2}{44.01 \text{ g}} = 0.050 \text{ mol } CO_2$$

Substitution of this number of moles, along with $V = 0.03236$ and $T = 298$ K into Equation 3 yields:

$$P = \frac{0.050 \text{ mol } (0.0821 \text{ L atm mol}^{-1}\text{K}^{-1})(298 \text{ K})}{0.03236 \text{ L}} = 3.8 \text{ atm}$$

This value is in agreement with pressures commonly found in soda bottles at room temperature.[1,2]

REFERENCES

1. Kuntzleman, T. S., & Sturgis, A. (2020). The effect of temperature in experiments involving carbonated beverages. *Journal of Chemical Education, 97*, 4033–4038.
2. https://www.youtube.com/watch?v=AlTsaFUqZsA

25

Real and Ideal Gases

TOPIC: real and ideal gases, gas laws

Materials:

- At least 1 L of liquid nitrogen in a Dewar.
- Cryogenic gloves and face shield.
- One balloon (the kind used to make balloon animals works best) filled with air, another filled with helium.

Hazards:

Liquid nitrogen is an extremely cold cryogenic liquid (−196 °C, 77 K). Contact with skin can cause severe frostbite or cold burns. Always handle liquid nitrogen using cryogenic gloves and safety goggles or a face shield. Use only containers designed for cryogenic liquids (such as Dewar flasks), and never seal liquid nitrogen in a closed system, as rapid vaporization can cause pressure buildup and explosive rupture.

Ensure the demonstration area is well ventilated, as nitrogen gas displaces oxygen and may pose an asphyxiation risk in poorly ventilated or confined spaces. Keep observers at a safe distance and instruct them not to touch any objects recently removed from the liquid nitrogen until they return to room temperature.

Procedure:

1. Simultaneously immerse both balloons into the liquid nitrogen.
2. Leaving the balloons immersed for 1 to 2 minutes, note any differences in the behavior of each balloon. Usually, the air-filled balloon shrinks to the point where it cannot reasonably be described as inflated. The helium balloon also shrinks but remains somewhat inflated.
3. Remove the balloons and place them on a table. Observe.

Discussion:

A balloon filled with air that is submerged in liquid nitrogen shrinks down to just a few mL in volume. This is because at such low temperatures, gas molecules lose kinetic energy, slow down significantly, and can therefore no longer keep the balloon inflated. Air condenses to liquid below 78 K, which is just above the temperature of liquid nitrogen (77 K). Therefore, when an air-filled balloon is submerged in liquid nitrogen, not only do the gas molecules move more slowly, but they are also chilled to below their condensation point. Because of this, the air molecules behave in a nonideal manner—that is, like a liquid. Thus, air at the temperature of liquid nitrogen does not behave ideally, but rather as a real gas.

A balloon filled with helium that is submerged in liquid nitrogen also decreases in volume, due to the helium atoms moving more slowly at 77 K. However, in this case the helium balloon remains somewhat inflated, shrinking down to about one-quarter of its original volume. This is because helium must be chilled to below 4 K to condense. At 77 K, the helium atoms still behave ideally (like a gas); therefore, the helium balloon remains somewhat inflated in the liquid nitrogen.

These observations can be made quantitative using Charles's law, which only applies to gases that behave ideally. Charles's law describes how the volume of a gas, V_1 (at temperature T_1), will change to volume, V_2, when cooled or warmed to temperature T_2:

$$\frac{V_1}{T_1} = \frac{V_2}{T_2} \tag{1}$$

At room temperature, gases may be estimated to have a temperature of 300 K. Assume a balloon initially with a volume of 500 mL at 300 K is chilled in liquid nitrogen to 77 K. Substituting these values into Equation 1 and solving for V_2, the volume of the balloon after being chilled we find:

$$V_2 = \frac{V_1}{T_1} T_2 = 128 \text{ mL}$$

Which is roughly one-quarter of the original volume of 500 mL. Charles's law accurately predicts the helium balloon's behavior, but fails for the air balloon due to condensation, which violates ideal gas assumptions

REFERENCE

1. Kuntzleman, T. (2015). Solution to chemical mystery #4: The case of the misbehaving balloon. *Chemical Education Xchange.* https://www.chemedx.org/blog/solution-chemical-mystery-4-case-misbehaving-balloon

26

Pressure in the Fire Syringe

TOPIC: combined gas law

Materials:

- Fire syringe (see https://www.arborsci.com/products/fire-syringe).

Hazards:

There are no significant hazards associated with this demonstration when the fire syringe is used as directed.

Procedure:

The fire syringe is a commercial product that is sold by several science supply stores. Suppliers typically provide instructions for proper use. Videos that demonstrate its use are available.[1]

1. Use a pair of tongs to insert a very tiny piece of cotton into the vessel in the fire syringe. Note that the internal volume in the vessel is approximately 10 mL.
2. Gently push down on the plunger of the syringe to observe the extent to which the volume can be compressed. In the opinion of the author, the volume may be decreased to about 0.5 mL.
3. Pull up the syringe, preparing to rapidly compress the plunger and ignite the cotton.
4. Push down very hard on the syringe, observing the contents of the vessel. If no ignition of the cotton occurs, troubleshoot and try again.

Discussion:

A piece of cotton placed in a vessel of air ignites when the pressure of the air is significantly increased. This occurs because the increased air pressure simultaneously causes an increase in the temperature of the air, and this heats the cotton to above its ignition temperature of 533 K.[2]

This demonstration may be made quantitative using the combined gas law. The combined gas law relates the pressures, *P*, volumes, *V*, and temperatures, *T*, of gases under some initial conditions (denoted with a 1 subscript) and final conditions (denoted with a 2 subscript).

$$\frac{P_1V_1}{T_1} = \frac{P_2V_2}{T_2} \tag{1}$$

Equation 1 can be used to estimate the pressure that must have been applied to the plunger to achieve the conditions that caused ignition. For our estimate, we first assume that the air in the syringe had an initial volume of 10 mL at 1 atm pressure and 300 K (room temperature). For ignition to occur, the gas in the syringe must have been at least heated to T_2 = 533 K. We further estimate the volume of the syringe was decreased to V_2 = 0.5 mL upon application of pressure to the plunger. Rearranging Equation 1, and inserting these values provides the estimate of the pressure that must have been applied to achieve these conditions:

$$P_2 = P_1\frac{T_2V_1}{T_1V_2} = 1\text{ atm}\frac{(533\text{ K})(10\text{ mL})}{(300\text{ K})(0.5\text{ mL})} = 36\text{ atm}$$

REFERENCES

1. Tommy Technetium. (2022). Using air pressure to burn cotton. https://www.youtube.com/watch?v=VNyX3CLBYIM
2. Graf, S. H. (1949). *Ignition temperatures of various papers, woods, and fabrics*. Engineering Experiment Station, Oregon State College.

27

Temperature of a Red-Hot Coin

TOPIC: heat equation, first law of thermodynamics

Materials:

- United States penny minted prior to 1982 (almost 100% copper).
- 250 mL beaker filled with ~75 mL of water precisely weighed.
- Tongs to hold hot penny.
- Bunsen burner or propane torch.
- Thermometer.

Hazards:

The penny will become very hot when heated and can cause burns. Always handle the penny using appropriate tongs or heat-resistant tools. Avoid touching the penny immediately after heating. Allow it to cool safely in the water. Use caution when working with open flames such as Bunsen burners or propane torches. Keep flammable materials away and follow standard laboratory safety procedures. Wear safety goggles and heat-resistant gloves as needed.

Procedure:

1. Find and record the mass of the copper penny (pre-1982). **Make certain that the penny is minted prior to 1982. Pennies minted on or after 1982 contain a large proportion of zinc which will melt in this experiment.**
2. Place ~75 mL of water in a beaker. Find and record the mass of the water.
3. Measure and record the initial temperature of the water.
4. Holding the penny with crucible tongs, heat the penny in a flame source until it glows.
5. Remove the penny from the flame and **IMMEDIATELY** drop the penny into the beaker of water.
6. Swirl the penny in the water. Measure the final temperature of the water once it stabilizes. This is assumed to be the final temperature of the penny.

Discussion:

This simple demonstration can be used to estimate the temperature of the penny when it was glowing hot. To do so, we first use the familiar heat equation:

$$Q = mc\Delta T \tag{1}$$

where Q is the energy gained as heat, m is mass of the water, c is specific heat (4.18 J g^{-1} °C^{-1} for water and 0.385 J g^{-1} °C^{-1} for copper), and ΔT is the change in temperature of the substance:

$$\Delta T = T_F - T \tag{2}$$

In Equation 2, T_F is the final temperature of the substance and T_I is the initial temperature of the substance. We now make the simplifying assumption that energy exchange only occurs between the penny and water. Therefore:

$$Q_{\text{water}} + Q_{\text{penny}} = 0 \tag{3}$$

Substitution of Equation 1 into Equation 3 yields:

$$m_{\text{water}}\, c_{\text{water}}\, \Delta T_{\text{water}} + m_{\text{penny}}\, c_{\text{penny}}\, \Delta T_{\text{penny}} = 0$$

Rearranging Equation 3, we can find the change in temperature of the penny:

$$\Delta T_{\text{penny}} = \frac{-m_{\text{water}}\, c_{\text{water}}\, \Delta T_{\text{water}}}{m_{\text{penny}}\, c_{\text{penny}}} \tag{4}$$

Insertion of Equation 2 into Equation 4 allows us to find the change in temperature of the penny:

$$\Delta T_{\text{penny}} = \frac{-m_{\text{water}}\, c_{\text{water}}\, \left(T_F - T_{I,\text{water}}\right)}{m_{\text{penny}}\, c_{\text{penny}}} \tag{5}$$

Finally, Insertion of Equation 2 into Equation 5 and rearranging gives us the equation for the temperature of the penny when it was hot:

$$T_{I,\text{penny}} = T_F + \frac{m_{\text{water}}\, c_{\text{water}}\, \left(T_F - T_{I,\text{water}}\right)}{m_{\text{penny}}\, c_{\text{penny}}} \tag{6}$$

The temperature of the penny found in this experiment usually falls below the melting temperature of copper (1085 °C). This is consistent with the observation that the copper penny does not melt during heating.

28

The Devil's Milkshake[1,2]

TOPICS: chemical thermodynamics, reaction enthalpy

Materials:

- 2 g of calcium metal.
- Phenolphthalein indicator (optional).
- 400 mL beakers half filled with water.
- Butane lighter with nozzle.
- Large and small Styrofoam cup (for quantitative demo).
- Balance (for quantitative demo).
- Thermometer.

Hazards:

Calcium metal is highly reactive, especially with water, and can produce flammable hydrogen gas and heat that may cause burns or ignite the hydrogen. Handle calcium metal with care using appropriate tools (e.g., tongs) and wear safety goggles and gloves. The flame produced when igniting hydrogen is often nearly invisible and can cause burns, so keep a safe distance and avoid inhaling any gases produced. Phenolphthalein is an irritant; avoid skin and eye contact. Perform this demonstration in a well-ventilated area or under a fume hood to prevent accumulation of hydrogen gas and ensure proper airflow.

Procedure:

1. Add a dropperful of phenolphthalein indicator to the water.
2. Add all of the calcium metal to the water.
3. Use the lighter to ignite the bubbles. Because the bubbles often extinguish, repeat ignition as necessary if desired.

Procedure for quantitative demonstration:

1. Add about 100 g of water—measured to the nearest gram—to the large Styrofoam cup.
2. Find and record the initial temperature of the water.
3. Add about 2 g of calcium metal—measured to the nearest milligram—to the large Styrofoam cup. After doing so, quickly cover the reaction in the large Styrofoam cup with the smaller Styrofoam (inverted) to minimize heat loss and trap the energy released in the reaction.
4. Once the reaction has subsided, find and record the final temperature of the water.

Discussion:

When calcium metal is added to water, numerous bubbles are produced, and a white, cloudy solid is produced within the water. If the bubbles are ignited, a sustained flame is observed on top of the mixture, and popping noises are heard. The formation of the cloudy-white solid gives the illusion that a "milkshake" is produced. If phenolphthalein indicator is added to the water, the indicator turns pink in color.

The chemical reaction involved in this experiment is the reaction between calcium metal and water to form insoluble calcium hydroxide and hydrogen gas:

$$Ca(s) + 2\,H_2O(l) \rightarrow Ca(OH)_2(s) + H_2(g) \qquad (1)$$

The bubbles that are formed contain hydrogen gas. When ignited, the hydrogen reacts with oxygen gas in the air to produce water and a lot of energy, producing the flame:

$$2\,H_2(g) + O_2(g) \rightarrow 2H_2O(g) \qquad (2)$$

If you look closely, you will note that the flame is an orange-red color, indicative of a positive test for the element calcium. The combination of the milky-white calcium hydroxide and red-orange flaming bubbles give this experiment the appearance of a flaming milkshake—The Devil's Milkshake! Of course one should never consume this "milkshake."

The Devil's Milkshake is made to have a "strawberry flavor" by adding phenolphthalein to the water prior to adding the calcium metal. A small amount of the $Ca(OH)_2$ that is formed dissolves, generating enough hydroxide ion to cause the phenolphthalein indicator to turn pink:

$$Ca(OH)_2(s) \rightarrow Ca^{2+}(aq) + 2\,OH^-(aq) \qquad K_{sp} = [Ca^{2+}][OH^-]^2 \qquad (3)$$

This is a fun experiment to do around Halloween time.

If desired, this demonstration can be used to estimate the enthalpy of the reaction outlined in Equation 1.[2] To do so, the energy released as heat, Q, during the reaction can be found using the following equation:

$$Q = mc\Delta T \qquad (4)$$

where m is the mass of water contained in the large Styrofoam cup, c is the specific heat of water (4.18 J g^{-1} °C^{-1}), and $\Delta T = T_F - T_I$, where T_I is the temperature of the water prior to the reaction and T_F is the temperature of the water at the completion of the reaction. The enthalpy of the reaction, ΔH, outlined in Equation 1 is then found by:

$$\Delta H = Q/n \qquad (5)$$

where n is the moles of calcium added to the water.

It is instructive to have students calculate the theoretical enthalpy of the reaction, , outlined in Equation 1, and compare that to the value experimentally determined in this demonstration. To do so, standard enthalpies of formation (ΔH_f^0,Table 1) and the following Equation are used:

$$\Delta H_{rxn} = \sum \Delta H^0_{f,\text{products}} - \Delta H^0_{f,\text{reactants}} \quad (6)$$

Table 1: *Standard enthalpies of formation of substances in Equation 1.*[3]

Substance	ΔH_f° (kJ mol^{-1})
Ca(*s*)	0
$H_2O(l)$	−286
$Ca(OH)_2(s)$	−986
$H_2(g)$	0

Insertion of the appropriate values from Table 1 into Equation 6 for the reaction outlined in Equation 1 yields:

$$\Delta H_{rxn} = -986 \text{ kJ mol}^{-1} + 0 - [0 - 2\,(286 \text{ kJ mol}^{-1})] = -414 \text{ kJ mol}^{-1}$$

Reference 2 outlines how to go through some of these calculations with experimental data.[2]

REFERENCES

1. Kuntzleman, T. (2019). The devil's milkshake. *Chemical Education Xchange.* https://www.chemedx.org/blog/devils-milkshake
2. Kuntzleman, T. (2020). Thermochemical analysis of the devil's milkshake. *Chemical Education Xchange.* https://www.chemedx.org/blog/thermochemical-analysis-devils-milkshake
3. U.S. Department of Commerce, National Institute of Standards and Technology. (2025). NIST Chemistry WebBook. https://webbook.nist.gov/chemistry/

29

Pharaoh's Serpent

TOPICS: chemical reactions, chemical thermodynamics, Gibbs energy

Materials:

- Hexamine fuel tablet.
- Calcium gluconate tablet.
- Lighter.

Hazards

This demonstration involves combustion and the release of hot gases and solid residues, so appropriate safety precautions must be taken. The hexamine fuel tablet is highly flammable and produces an open flame during combustion. Care should be taken to ignite it only on a heat-resistant, nonflammable surface such as a brick or metal tray, and away from flammable materials.

Calcium gluconate decomposes at high temperature, producing hot carbonaceous foam that can cause burns on contact. The foam is fragile and may break apart, releasing hot particles and gases.

The combustion products include carbon dioxide, water vapor, and possibly small amounts of carbon monoxide and other combustion byproducts—adequate ventilation is required to prevent accumulation of gases in enclosed spaces.

Eye protection (goggles) and heat-resistant gloves should be worn during the demonstration.

Keep a fire extinguisher or bucket of sand nearby in case of accidental fire spread.

Perform the experiment under a fume hood or well-ventilated area and never inhale smoke or vapors directly. Due to the open flame and hot products, maintain a safe distance and do not touch the materials or products until fully cooled.

Procedure:

1. Place the fuel tablet on a flat, fireproof surface (a brick works well).
2. Place the calcium gluconate tablet on top of the fuel tablet.
3. Ignite the fuel tablet.
4. Observe.

Discussion:

In this experiment, calcium gluconate thermally decomposes according to:[1,2]

$$CaC_{12}H_{22}O_{14}(s) \rightarrow 11\ C(s) + 11\ H_2O(g) + CaCO_3(s) \qquad (1)$$

This process displays several interesting phenomena. First, a column of black foam that looks like a snake is formed. The color of the column changes to gray and then to white as the reaction proceeds. The foam "snake" glows orange when it is still black but shifts to emitting white light its color changes to gray or white. It is the production of carbon (Equation 1) that gives the foam its original black color. When heated to high temperatures, carbon glows an incandescent orange, so this explains the orange emission. The black foam shifts to a white color because the carbon is converted to CO_2:

$$11\ C(s) + 11\ O_2(g) \rightarrow 11\ CO_2(g) \qquad (2)$$

The black carbon is thus removed, but the white $CaCO_3$ formed in Equation 1 remains behind. Thus, the foam shifts to a white color.

Both gaseous water (Equation 1) and carbon dioxide (Equation 2) permeate the solid products to cause the foam.

The intense white light emitted from the foam when it is white occurs because $CaCO_3$ is converted into CaO:

$$CaCO_3(s) \rightarrow CaO(s) + CO_2(g) \quad (3)$$

When the CaO product, which is also known as lime, is heated to high temperatures, it emits bright white light: This is the origin of the term "limelight."

Adding together Equations 1-3 yields the following overall reaction for the reaction:

$$CaC_{12}H_{22}O_{14}(s) + 11\ O_2(g) \rightarrow 12\ CO_2(g) + 11\ H_2O(g) + CaO(s) \quad (4)$$

Various thermodynamic data (Table 1) may be used to quantitatively explore Equations 1-3 in more detail. To do so, the following equations are used:

$$\Delta H^o_{rxn} = \Delta H^o_{f,products} - \Delta H^o_{f,reactants} \quad (5)$$

$$\Delta S^o_{rxn} = S^o_{products} - S^o_{reactants} \quad (6)$$

$$\Delta G = \Delta H - T\Delta S \quad (7)$$

Using Equations 5-7 and the values in Table 1, the values of ΔH^o_{rxn}, ΔS^o_{rxn}, and ΔG^o_{rxn} for reactions 1 to 3 displayed in Table 2 are obtained.

Table 1: Thermodynamic data of the species in Equations 1 to 5[3-5]

Compound	ΔH_f^o (kJ mol^{-1})	S^o (J mol^{-1} K^{-1})
$CaC_{12}H_{22}O_{14}$	−3545	472
$CaO(s)$	−635	38
$CaCO_3(s)$	−1207	93
$CO_2(g)$	−394	214
C(*s*, amorphous)	20	20
$H_2O(g)$	−242	189
$O_2(g)$	0	205

Table 2: Thermodynamic values for Equations 1 to 3 at $T = 298.15$ K

Equation	ΔH_{rxn}^{o} (kJ mol^{-1})	S^{o} (J mol^{-1} K^{-1})	ΔG_{rxn}^{o} (kJ mol^{-1})
1	−104	+1920	−676
2	−374	−11	−371
3	+178	+159	+131

Interestingly, ΔG is positive for Equation 3 at 298 K, indicating non-spontaneity. However, the reaction does become spontaneous at higher temperature. The temperature at which Equation 3 becomes spontaneous can be estimated by insertion of ΔH^{o} (= +178,000 J mol^{-1}) and ΔS^{o} (+159 J mol^{-1} K^{-1}) for Equation 3 into Equation 8:

$$T \approx \Delta H / \Delta S \qquad (8)$$

Upon doing so, a value of $T = 1120$ K is obtained. Indeed, both Equations 1 and 2, as well as the burning of the fuel tablet:[5]

$$C_6H_{12}N_4(s) + 9\,O_2(g) \rightarrow 6\,CO_2(g) + 2\,N_2(g) + 6\,H_2O(g) \qquad (9)$$

all together account for heating the foam to a high enough temperature to allow Equation 3 to become spontaneous. Further, this heating allows the CaO product to emit limelight!

REFERENCES

1. MEL Science. (n.d.). "Gluconate snake" experiment. https://melscience.com/US-en/articles/gluconate-snake-experiment/
2. Labuschagne, F. J. W. J., & Focke, W. W. (2003). Metal catalysed intumescence: Characterisation of the thermal decomposition of calcium gluconate monohydrate. *Journal of Material Science, 38*, 1249–1254.
3. Di, Y.-Y., Zhang, G.-C., Lui, Y.-P., Kong, Y.-X., & Zhou, C.-S. (2021). *Journal of Molecular Structures, 1225*, 1–10.
4. Eisermann, W., Johnson, P., & Conger, W. L. (1980). Estimating thermodynamic properties of coal, char, tar and ash. *Fuel Processing Technology, 3*, 39–53.
5. Ebbing, D. D., & Gammon, S. D. (2017). *General chemistry* (11th ed.). Cengage Learning.
6. Merritt, J. R., Herington, L., Jones, S. B., & Sayed, Y. (1991). *American Industrial Hygiene Association Journal, 52*, 30–33.
7. Kuntzleman, T. (2024). The chemistry of Pharaoh's serpent with calcium gluconate. *Chemical Education Xchange*. https://www.chemedx.org/blog/chemistry-pharaohs-serpent-calcium-gluconate

30

Enthalpy of Sublimation of Solid Carbon Dioxide

TOPICS: chemical thermodynamics, heat equation, enthalpy of sublimation

Materials:

- Large Styrofoam cup.
- Smaller Styrofoam cup that can be nested inside the larger cup.
- Dry ice pellets.
- Balance.
- Thermometer.
- Water.

Hazards:

Dry ice is extremely cold, with a temperature of about −78.5 °C. Direct contact with dry ice can cause severe cold burns or frostbite on skin and other tissues. Always handle dry ice with appropriate personal protective equipment, such as insulated gloves or tongs. Avoid placing dry ice in sealed containers, as sublimation produces carbon dioxide gas which can build up pressure and cause the container to rupture. Additionally, dry ice releases carbon dioxide gas, which can accumulate in poorly ventilated areas and cause asphyxiation. Perform this demonstration in a well-ventilated area and never inhale carbon dioxide gas directly.

Procedure:

1. Place about 200 g of water weighed to the nearest gram in the large Styrofoam cup.
2. Find and record the temperature of the water. Preheating the water to ~50°C improves experimental results.[1]
3. Place about 15 g of dry ice weighed to the nearest tenth of a gram in the water. Convert this mass to moles of CO_2 and record. **NOTE:** Crushing dry ice with a hammer can help reduce pellet size but makes accurate weighing more difficult.[1]
4. Immediately, invert the smaller cup and use it to cover the large Styrofoam cup.
5. Once the dry ice has completed subliming, find and record the final temperature of the water.
6. Use Equation 5 (see Discussion) to determine the enthalpy of sublimation of CO_2.

Discussion:

Dry ice, which is composed of solid carbon dioxide, undergoes sublimation. In other words, it transforms directly from the solid phase to the gas phase:

$$CO_2(s) \rightarrow CO_2(g) \qquad \Delta H = +25 \text{ kJ mol}^{-1} \tag{1}$$

Molecules in the solid phase need to gain kinetic energy to transition to the gas phase. Consistent with this fact, the enthalpy of sublimation of dry ice (ΔH) is positive for Equation 1.[1,2]

This demonstration illustrates how to estimate ΔH for this reaction. To do this, a lump of dry ice of known mass is placed in a known mass of water at a known temperature. As the solid CO_2 sublimes it gains energy from the water, causing the temperature of the water to drop. The energy lost as heat by the water, Q_w, can be determined by:

$$Q_w = m_w c \Delta T \tag{2}$$

where m_w is the mass of water in the cup, $c = 4.18 \text{ J g}^{-1} \text{ °C}^{-1}$ is the specific heat of water, and ΔT is the change in temperature that occurred in the water, or:

$$\Delta T = T_F - T_I \tag{3}$$

where T_F and T_I are the final and initial temperatures of the water, respectively. From the first law of thermodynamics we know that the energy lost by the water was gained by the dry ice. In other words:

$$-Q_w = n\Delta H \tag{4}$$

where n is the moles of CO_2 that sublimed during the demonstration. Substitution of Equations 2 and 3 into Equation 4, and then dividing by n yields the molar enthalpy of sublimation of CO_2 in terms of values experimentally measured in this demonstration:

$$\Delta H = \frac{m_w c(T_F - T_I)}{n} \tag{5}$$

References

1. Burgstahler, B. (1991). *Journal of Chemical Education, 68*, 331–333.
2. National Institute of Standards and Technology. (n.d.). Phase change data. https://webbook.nist.gov/cgi/cbook.cgi?ID=C124389&Mask=4#Thermo-Phase

31

Wavelength of Light

TOPICS: measurement of length, unit conversion, wave nature of light

Materials:

- Laser pointers with different wavelengths: 650 nm, 532 nm, 405 nm.
- Diffraction glasses with a known grating spacing.[1]
- Meterstick.
- A flat surface (screen) onto which you will project the laser light.

Hazards:

Laser pointers can cause serious eye injury if their beams are viewed directly or reflected into the eyes. Never look directly into a laser beam or point the laser at anyone's eyes.

Procedure:

1. Position yourself at least 3 m from the screen.
2. Confirm the grating spacing, *d*, of your diffraction glasses before starting
3. Measure the distance from the screen to the place where you will hold the diffraction grating. Call this distance *L*.
4. Shine the red laser light through the diffraction grating and onto the screen.
5. With the help of an assistant from the audience, measure the distance between the center dot of laser light on the screen to spot of light that is closest to this center dot. Call this distance *y*, the distance between the central bright spot and the first-order diffracted spot.
6. Use Equation 1 to calculate the wavelength, λ, of laser light: based on your measurements:

$$\lambda = d\frac{y}{\sqrt{L^2+y^2}} \tag{1}$$

   where λ is the wavelength of the laser light, *d* is the distance between spacings in the diffraction grating, *L* is the distance between the screen and diffraction grating, and *y* is the distance between the center dot of light and the spot of light closest to it.

7. Convert the wavelength measured to units of nanometers (nm).
8. If desired, repeat the experiment with the green and violet lasers.

Discussion:

When light passes through a diffraction grating and onto a screen, some of the light is diffracted (bent) from its straight-line path (Figure 1). Most light passes straight through the grating, forming a bright central spot on the screen, while some is diffracted to form additional spots. The light that is bent is observed as a series of bright spots that appear at regular intervals from the central spot.

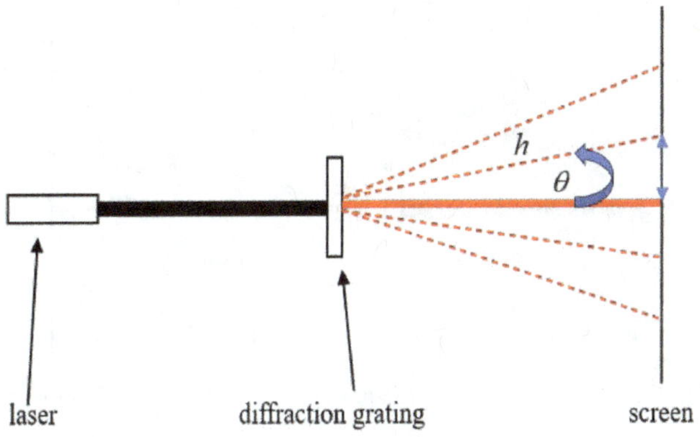

FIGURE 1: Laser light passing through a diffraction grating onto a screen. Most light passes straight through the grating (bold red line). Some light (dashed red lines) is diverted. The distance from the grating to the screen is designated L. The blue arrow designates the distance between the central bright spot of light and the spot of light closest to it.

A mathematical relationship exists between the observed distance of a spot of light nearest to the central bright spot (y), the wavelength of light (λ), the length from the diffraction grating to the screen (L) and the distance between slits within the diffraction grating (d) This relationship is:

$$\lambda = d\frac{y}{\sqrt{L^2+y^2}} \tag{1}$$

Equation 1 can be derived using the following equation for the diffraction of light through a diffraction grating:

$$\lambda = d\sin\theta \tag{2}$$

where λ is the wavelength of light, d is the distance between slits in the diffraction grating, and θ is the angle made between the straight-line beam of light and its next nearest neighbor. Notice that the diffracted beam makes up the hypotenuse, h, of a triangle with legs that have distances of L and y (Figure 2).

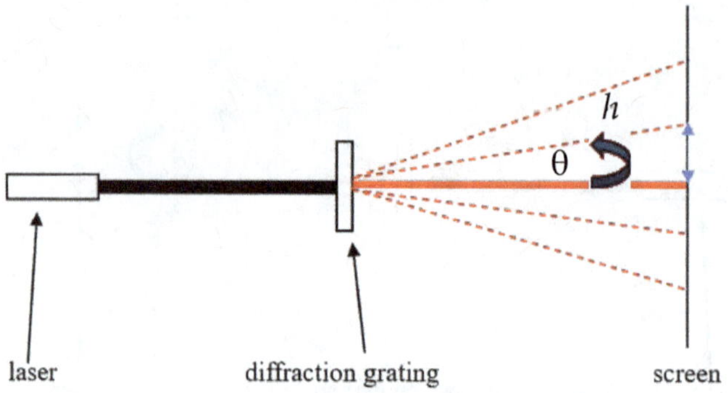

FIGURE 2: The angle between the straight-line beam and diffracted beam is θ. The hypotenuse of the triangle formed is designated h.

If we substitute sinθ = y/h into Equation 2, we obtain:

$$\lambda = d\frac{y}{h} \tag{3}$$

By the Pythagorean theorem, $h^2 = L^2 + y^2$. Substitution of this result into Equation 3 gives us the equation we seek:

$$\lambda = d\frac{y}{\sqrt{L^2+y^2}} \tag{4}$$

This demonstration finds use in a variety of contexts. The analysis of data requires a fair bit of measurement and metric unit conversion, so it fits well when covering these topics. In addition, this demonstration touches upon the wave nature of light. As a result, it is a nice experiment to perform when introducing quantum chemistry.

Note:

1. Using the distance between the slits (d) in the diffraction grating as recorded by the manufacturer may cause some trouble in this experiment. For example, the author has used diffraction glasses listed as having 500 lines per mm, which would imply d = 2000 nm (1 mm /500 lines = 0.002 mm = 2000 nm. However, observing the grating with an optical microscope fitted with a length scale allowed for a direct measurement of d = 4850 nm. Thus, if it is noticed that measured wavelengths don't make sense (200 nm for red light, for example), consider measuring d with an optical microscope. If an optical microscope fitted with a length scale is not available, then simply conduct this experiment with light of known wavelength and use the following equation to determine d:

$$d = \lambda \frac{\sqrt{L^2+y^2}}{y} \quad (5)$$

REFERENCE

1. Kuntzleman, T. (2018). A simple method for determining the wavelength of light. *Chemical Education Xchange*. https://www.chemedx.org/blog/simple-method-measure-wavelength-light

32

Safe Flame Tests

TOPIC: flame tests

Materials:

- Mr. Clean Magic Eraser sponge.
- Scissors.
- Methanol or ethanol.
- $CuCl_2$, NaCl, LiCl, KCl, boric acid.
- Five to six wide-mouth containers, each large enough to hold a 1 in cube. Each container should have a tight-fitting stopper.
- Butane lighter with nozzle.
- Tongs.
- 1 L beaker half filled with water.

Hazards:

MAKE CERTAIN THAT ALL ALCOHOL CONTAINERS ARE STOPPERED TIGHTLY DURING IGNITION. ALSO KEEP ALCOHOL CONTAINERS STOPPERED WHEN FLAME TESTS ARE BEING CONDUCTED. DO NOT ATTEMPT TO POUR ALCOHOL ONTO A FLAME THAT IS "DYING OUT".

Procedure:

1. Fill each container half full with alcohol.
2. Keep one container filled with alcohol only—do not add any solid.
3. In the other containers, separately dissolve some $CuCl_2$, NaCl, LiCl, KCl, $SrCl_2$, and boric acid. It's OK if some undissolved solid remains on the bottom.
4. Cut the Mr. Clean Magic Eraser sponge into ½ to ¾ in cubes.
5. Immerse a sponge cube into each container **and stopper tightly**.
6. Using tongs, remove the sponge from the container with alcohol only (no dissolved solid). While removing, squeeze to remove excess liquid.

CAUTION: RESTOPPER THE CONTAINER AND REMOVE IT FROM THE DEMONSTRATION AREA BEFORE PROCEEDING.

7. Using tongs, hold the sponge at arm's length over the beaker filled with water and ignite it.
8. Observe the color of the flame. Note that this is the "control" color.
9. Extinguish the flame by plunging it into the beaker of water.
10. Repeat steps 6 to 9 for the other containers that contain various dissolved substances. Note the color each time.

Discussion:

When ignited, the flame surrounding each sponge will display the color associated with a positive flame test for the metal contained in the salt (**lithium = scarlet, sodium = yellow, potassium = lavender-orange, copper = green**). In the case of the metal salts, the emitted colors arise from electron transitions between quantized energy levels in the metal ions. In the case of the alcohol alone, a blue flame is observed. This blue flame is characteristic of complete combustion of carbon-containing materials. The blue color of the alcohol flame arises from excited carbon-containing species produced during combustion, making it a useful diagnostic indicator of carbon-containing materials burning. When the alcoholic solution of boric acid is ignited, a green flame is observed. This green color is commonly observed when compounds containing boron are burned.

REFERENCE

1. Landis, A. M., Davies, M. I., Landis, L., & Thomas, N. C. (2009). Magic eraser flame tests. *Journal of Chemical Education, 86*, 577–578.

33

Curie Temperature of Nickel

TOPIC: electronic configurations

Materials:

- Canadian nickel (5 cent piece) minted between 1955 and 1981 (inclusive), or Canadian quarters (25 cent piece) minted between 1968 and 1999 (inclusive).
- 250 mL beaker and water.
- Propane torch or Bunsen burner.
- Steel crucible tongs.
- Neodymium magnet.

Hazards:

Hot coins present a burn hazard. When heating the coin it must be held above a beaker of water so it becomes trapped in the beaker of water when it falls. Tongs become extremely hot when heating; handle with care. Tongs should be immersed in water immediately after heating. Neodymium magnets present a pinch hazard; use care to avoid pinching fingers. Tongs become hot on heating, so immerse tongs in water immediately after use.

Procedure:

A short video that displays how this experiment is carried out can be viewed at the link below:

https://youtube.com/shorts/lrpOc6IGojI

1. Obtain a Canadian coin made of the element nickel (see Materials).
2. Obtain a 250 mL beaker.
3. Place about 100 mL of water in the beaker.
4. Using caution not to pinch your fingers, attach a neodymium magnet to the steel tongs.
5. Attach the coin to the magnetized tongs. The coin should stick to the tongs but hang freely.
6. Use the magnetized tongs to hold the coin above the water.
7. Heat the coin with a propane torch until it loses its magnetic attraction and drops into the water.
8. After removing the coin from the water, the coin can again be attracted to the magnetized steel tongs.

Discussion:

While we cannot directly "see" the electronic configuration of nickel atoms, this demonstration shows a physical property—magnetism—that arises from the electron arrangement, thus providing an accessible way to connect abstract electronic structure concepts to observable phenomena.

The Curie temperature (T_C) is the point above which a substance loses its permanent magnetic properties. The element nickel is a ferromagnetic substance with a $T_C = 627$ K. Thus, at room temperature nickel is attracted to a pair of magnetized steel tongs. However, nickel metal can be heated to above its Curie temperature, causing it to no longer be magnetically attracted to the magnetized steel tongs.

The magnetic properties of nickel can be qualitatively connected to its electronic configuration. As a result, it is useful to carry out this demonstration when discussing the electronic configurations of the elements. The electronic configuration of nickel is $[Ar]3d^84s^2$, and according to Hund's rule nickel atoms have two unpaired electrons (Figure 1, top). Unpaired electrons are a requirement for an element to display paramagnetic or ferromagnetic behavior, and nickel clearly displays ferromagnetic behavior in this demonstration. Unfortunately, students tend to forget to apply Hund's rule when drawing orbital box diagrams of the elements. Ignoring Hund's rule when constructing the orbital box diagram for nickel can lead to the incorrect conclusion that nickel has all of its electrons paired—that is, no unpaired electrons (Figure 1, bottom, for example)

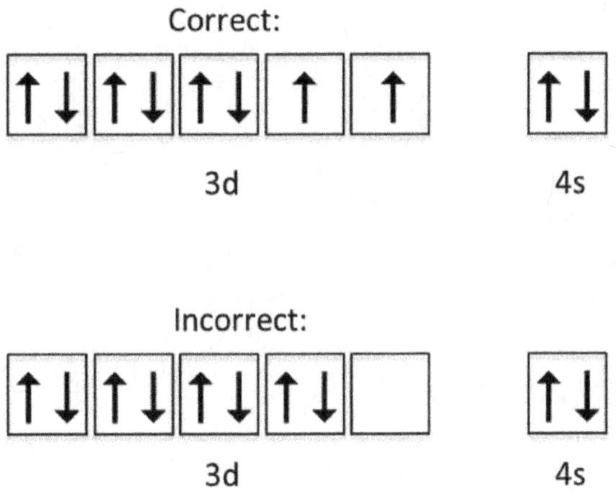

FIGURE 1: Correct (top) and a commonly drawn incorrect (bottom) orbital box diagram of the 3d and 4s subshells of nickel.

Thus, this demonstration can serve as a reminder for the importance of applying Hund's rule to obtain the correct electronic configuration for nickel, and for the other elements.

REFERENCE

1. Kuntzleman, T. S. (2023). Estimation of the Curie temperature of nickel using a simple thermodynamic analysis. *Journal of Chemical Education, 100,* 2789–2792.

34

The Chemistry of Lightning

TOPICS: chemical reactions, chemical thermodynamics, chemical spontaneity, Gibbs energy, general interest

Materials:

- Oudin coil.
- Petri dish.
- A universal indicator.
- Water.
- 0.1 M NaOH.

Hazards:

This demonstration involves the use of an Oudin coil, which produces high-voltage sparks that can cause electric shock or burns; therefore, care should be taken to avoid direct contact. The 0.1 M sodium hydroxide solution is caustic and can cause irritation to the skin and eyes, so appropriate protective gloves and eye protection should be worn when handling it. Although the universal indicator is generally low hazard, it should still be kept away from the eyes and mouth. Additionally, because sparks are produced, the demonstration area must be clear of any flammable materials to prevent accidental ignition.

Procedure:

A short video that displays how this experiment is carried out can be viewed at the link below:

https://www.youtube.com/shorts/kvtS7Fx9s8w

1. Add water to the petri dish.
2. Add the universal indicator to the water.
3. Add one drop of 0.1 M NaOH. If the indicator does not turn purple, continue adding 0.1 M NaOH, one drop at a time, until the indicator turns purple.
4. Turn on the Oudin coil and allow the sparks to hit the surface of the water. Continue to do this, noting the color changes that occur. It will take some time, but you should notice the indicator changing color as the pH decreases.

Discussion:

This demonstration shows that sparks from an Oudin coil chemically modifies gases in the air which, when dissolved in water, cause a decrease in pH. This effect is analogous to what occurs when lightning strikes water in the environment. Lightning exhibits some of the most fascinating phenomena on planet Earth. These massive strikes involve electrical potentials of 300 million volts[1] at 300,000 Amps and can generate temperatures up to 30,000 K.[2–4] In fact, lightning is known to emit both X-rays and gamma rays, and even kicks off some nuclear reactions.[2,3,5] It's fun to imagine all the amazing and exotic chemical reactions that might be possible under such drastic conditions.

The atmosphere contains mostly nitrogen and oxygen gases, so it's clear that lightning can produce nitrogen oxides (NO_x) from air components:[2–7]

$$N_2(g) + O_2(g) \rightarrow 2\ NO(g) \quad (1)$$

$$2\,NO(g) + O_2(g) \rightarrow 2\,NO_2(g) \qquad (2)$$

These reactions are not thermodynamically favorable under the conditions normally present in the atmosphere. But lightning strikes provide more than enough energy to drive these reactions to completion. In fact, it has been estimated that between 33 and 660 moles of NO are produced with every flash of lightning.[1,2] Considering that 9 million lightning strikes occur around the world every day,[1] lightning bolts account for as many as 6 billion moles of NO produced daily!

The NO_x produced by lightning flashes further react with available water to produce nitrous and nitric acids (Equation 3):[8]

$$2\,NO_2(g) + H_2O \rightarrow HNO_2(aq) + HNO_3(aq) \qquad (3)$$

As a result of all these processes, lightning strikes move nitrogen from the atmosphere into the soil.

It's possible to connect Equations 1 and 2 to topics in chemical thermodynamics. For example, using standard thermodynamic values (Table 1), standard enthalpies, entropies, and Gibbs energies of the processes in Equations 1 and 2 can be determined (Table 2).

Table 1: Thermodynamic values for select compounds

Compound	**ΔH_f° (kJ mol^{-1})**	**S° (J mol^{-1} K^{-1})**	**ΔG_f° (kJ mol^{-1})**
$N_2(g)$	0	191.6	0
$NO(g)$	90.3	210.8	86.6
$NO_2(g)$	33.1	240.0	51.2
$O_2(g)$	0	205.1	0

Table 2: Thermodynamic values for processes described in Equations 1 and 2

Reaction	**ΔH° (kJ mol^{-1})**	**ΔS° (J mol^{-1} K^{-1})**	**ΔG° (kJ mol^{-1})**
$N_2(g) + O_2(g) \rightarrow 2\,NO(g)$	+180.6	+24.9	+173.2
$2\,NO(g) + O_2(g) \rightarrow 2\,NO_2(g)$	−114.4	−146.7	−70.8

It's interesting to note that the processes described by Equations 1 and 2 are thermodynamic opposites of each other in many ways. For example, the process in Equation 1 is not spontaneous under standard conditions (ΔG is positive), but the reaction described by Equation 2 is (ΔG is negative). Furthermore, the formation of NO from nitrogen and oxygen (Equation 1) is not favored enthalpically (ΔH is positive) but is favored entropically (ΔS is positive). This situation is reversed for the formation of NO_2 (Equation 2), which is favored enthalpically (ΔH is negative) but not entropically (ΔS is negative).

Using the equation:

$$\Delta G = \Delta H - T\Delta S \tag{4}$$

it is seen that the formation of NO (Equation 1) is predicted to be spontaneous at higher temperatures but not lower ones. The temperature at which NO formation (Equation 1) becomes spontaneous can be estimated to be about 7250 K using the following equation and the values for ΔH and ΔS displayed in Table 2:

$$T \approx \Delta H^{o}/\Delta S^{o} \tag{5}$$

Therefore, while the formation of NO (Equation 1) is nonspontaneous at lower temperatures, it becomes spontaneous under the high temperature conditions associated with a lightning strike.

Using Equation 4 it can also be seen that the formation of NO_2 (Equation 2) is predicted to be spontaneous at lower temperatures, but not higher ones. If we again use Equation 5 and the appropriate values in Table 2, it is seen that the formation of NO_2 becomes nonspontaneous at temperatures that exceed roughly 780 K. According to this estimated temperature, the formation of NO_2 likely proceeds after the air cools slightly following a blast of lightning. Overall, this analysis would predict that the initial blast probably generates high concentrations of NO (Equation 1) which then react with atmospheric oxygen (Equation 2) once the air cools. It should be kept in mind that these are simple thermodynamic estimates that ignore concentration effects and the incredibly high electrical potentials associated with a lightning strike. Nevertheless, it has been suggested that NO forms simultaneously with lightning strikes and NO_2 forms as the surrounding air cools thereafter.[10]

NOTES AND REFERENCES

1. Price, C., & Penner, J. (1997). NO_x from lightning. *Journal of Geophysical Research, 102*, 5929–5941.
2. Dwyer, J. R., & Uman, M. A. (2014). The physics of lightning. *Physics Reports, 534*, 147–241.
3. Gibb, B. C. (2019). Lightning-fast chemistry. *Nature Chemistry, 11*, 677–679.
4. Hill, R. D., Rinker, R. G., & Wilson, H. D. (1980). Atmospheric nitrogen fixation by lightning. *Journal of Atmospheric Science, 37*, 179–192.
5. Enoto, T., Wada, Y., Furuta, Y., et al. (2017). Photonuclear reactions triggered by lightning discharge. *Nature, 551*, 481–484.
6. Griffing, G. W. (1977). Ozone and oxides of nitrogen production during thunderstorms. *Journal of Geophysical Research, 82*, 943–950.
7. Although not covered here, lightning bolts also generate ozone gas ($3\ O_2 \rightarrow 2\ O_3$), and ozone gas can react with NO to produce NO_2. See references 1, 4, and 6.
8. Zhu, R. S., Lai, K.-Y., & Lin, M. C. (2012). Ab initio chemical kinetics for the hydrolysis of N_2O_4 isomers in the gas phase. *Journal of Physical Chemistry A, 116*, 4466–4472.
9. For more experiments connecting lightning and atmospheric components, see Wagner, E. P., & Murray, B. E. (2021). An undergraduate experiment to investigate air pollutants created from lightning. *Journal of Chemical Education, 98*, 1361–1370.
10. Karabulut, E. (2022). *Journal of Physical Chemistry A, 126*, 5363–5374.
11. Kuntzleman, T. (2022). The chemistry of lightning. *Chemical Education Xchange*. https://www.chemedx.org/blog/chemistry-lightning

NOTES AND REFERENCES

[illegible]

35

Go Blue!

TOPIC: intermolecular forces

Materials:

- 500 mL graduated cylinder with a tight-fitting stopper.
- 200 mL of water.
- 200 mL of 70% isopropyl alcohol.
- 50 g of potassium carbonate (sodium carbonate will work, but the water and alcohol need to be heated).
- Green food dye that contains blue #1 (blue 1) and yellow #5 (yellow 5). Alternatively, blue food dye containing blue #1 and yellow food dye containing yellow #5 can be used.

Hazards:

Isopropyl alcohol is flammable. Keep from flame.

Procedure:

1. Pour the 70% alcohol and water into the graduated cylinder. Stopper and mix well.
2. To this mixture, add five to seven drops of the green food dye (or add three drops each of blue and yellow food dye). Stopper and mix well.
3. Add the potassium carbonate (white) to the green mixture, with a nod to Michigan State fans, who may appreciate the green-and-white motif. Stopper and mix well.
4. Place the cylinder in an area where all can observe for 1 minute.
5. The contents of the cylinder separate into a yellow layer on bottom and a blue layer on top: the colors of the University of Michigan—Michigan State's main rival!

Disposal:

The bottle may be saved for future viewing/demonstrations.

Explanation:

Green food dye contains a mixture of blue 1 and yellow 5 (Figure 1). It is instructive to note that blue 1 has an overall charge of −2 and contains 11 hydrogen bond acceptors (N or O atoms), yellow 5 has an overall charge of −3 and contains 13 hydrogen bond acceptors. These properties help to explain why these dyes are very soluble in water.

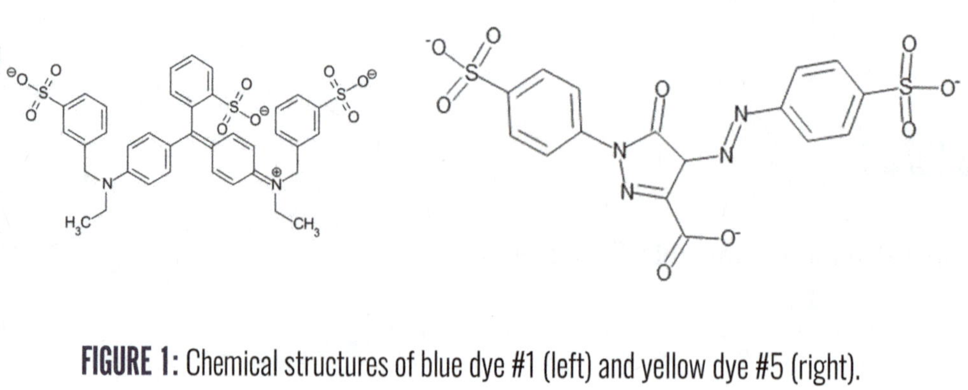

FIGURE 1: Chemical structures of blue dye #1 (left) and yellow dye #5 (right).

These components readily mix to form a homogeneous solution. There is a rich assortment of intermolecular forces that are responsible for this mixing. The water and alcohol mix because of dipole–dipole and hydrogen bonding interactions; the dyes mix with the alcohol and water due to dipole–dipole, hydrogen bonding, and ion–dipole interactions.

When potassium carbonate is added to the green mixture, very strong ion–dipole forces are set up between water molecules and potassium ions, as well as between water molecules and carbonate ions. These strong ion–dipole forces outcompete the intermolecular forces between the molecules of water and isopropyl alcohol. As a result, the alcohol molecules become “squeezed out”. This results in the mixture separating into two layers: a less dense, alcohol rich layer on top and a more dense, aqueous salt layer on bottom.

The yellow dye has a greater overall charge and more hydrogen bond acceptors than the blue dye. Because of these factors, the yellow dye is able to strongly interact with the molecules and ions in the aqueous layer. Thus, it remains in the water-rich layer. The blue dye, with a lower charge and fewer hydrogen bond acceptors is not strongly attracted enough to water molecules to remain in the aqueous layer. Therefore, it separates into the alcohol-rich layer.

REFERENCES

1. Kuntzleman, T. (2016). Chemical mystery #8: Go blue! *Chemical Education Xchange.* https://www.chemedx.org/blog/chemical-mystery-8-go-blue
2. Kuntzleman, T. (2016). Solution to chemical mystery #8: Go blue! *Chemical Education Xchange.* https://www.chemedx.org/blog/solution-chemical-mystery-8-go-blue
3. Kuntzleman, T. (2024). Exploring color separating in salting out experiments. *Chemical Education Xchange.* https://www.chemedx.org/blog/exploring-color-separation-salting-out-experiments

36

Food Dye Kinetics

TOPIC: chemical kinetics

Materials:

- Green food dye that contains Blue Dye #1 and Yellow Dye #5.
- Deionized water that has been heated.
- Deionized water at room temperature.
- 600 mL beaker.
- 4 × 500 mL beakers.
- Arm & Hammer Super Washing Soda (Na_2CO_3).
- 3% hydrogen peroxide.

Hazards:

Solutions of sodium carbonate are basic. Wear gloves and goggles.

Procedure:

1. Add 500 mL of water to the beaker 600 mL.
2. Add 1 TBSP (15 mL) of washing soda and 1 drop of green food coloring to the water. Stir until dissolved.
3. Divide the green solution equally between two 500 mL beakers. Label them D1 and D2 (for dye).
4. In a third 500 mL beaker, place ½ TBSP (7.5 mL) of 3% hydrogen peroxide. Label this beaker H1.
5. In the fourth 500 mL beaker, place 1 TBSP (15 mL) of 3% hydrogen peroxide. Label this beaker H2.
6. At the same time, pour D1 into H1, and D2 into H2. Observe.
7. Note any color changes. If desired, time how long it takes for the green color to fade to yellow in each beaker.
8. If desired, repeat this experiment using water at a higher temperature.

Disposal:

Place solutions in a waste container indicated for aqueous base.

Discussion:

Generally speaking, green food dye contains a mixture of Blue Dye #1 and a yellow dye (usually Yellow #5). In the presence of base, Blue dye #1 is oxidized by hydrogen peroxide to form colorless products:

$$\text{Blue \#1} + H_2O_2 \rightarrow \text{Colorless products} \qquad (1)\ (\text{in the presence of } OH^-)$$

On the other hand, Yellow #5 is not oxidized under these conditions. When hydrogen peroxide and green dye are mixed in the presence of Na_2CO_3 (which is basic), the green dye appears to fade to a yellow color. This occurs because the Blue #1 in the green dye reacts according to Equation 1, leaving behind only the yellow dye.

In these experiments, the concentration of H_2O_2 is varied between the two beakers, with twice as much in one case as compared to the other. The fading color is observed to happen more quickly in the beaker with the higher concentration of H_2O_2. Therefore, the rate of the reaction outlined in Equation 1 depends upon the concentration of H_2O_2. Usually, the rate of this reaction is found to vary linearly with the concentration of H_2O_2. In other words, the reaction rate approximately doubles when the concentration is doubled, indicating first order with respect to H_2O_2.

This reaction takes longer if the experiment is conducted in water at room temperature as compared to hot water. This is consistent with the fact that most chemical reactions occur more quickly at higher temperatures.

Thus, this demonstration illustrates two facets of chemical kinetics: 1) Reaction rates depend on the concentration of reactants, and 2) reaction rates depend upon temperature. If it is desired to show the effect of a catalyst on a reaction, this experiment may be set up and run in the absence of Na_2CO_3. Upon doing so, the reaction does not proceed to any observable extent within several minutes.

37

Can Alkaline Water Change Your Body's pH?

TOPIC: concentration of acids, strength of acids, K_a and K_b, pH, acid–base titrations

Materials:

- One bottle each of Aquafina (any size), Fiji (700 mL), and Essentia (700 mL) waters. Others types may also be purchased for use.
- A universal indicator.
- 3 M HCl.
- Plastic droppers. For more precise work, a micropipette capable of delivering 366 μL of fluid works best.
- A diagram showing the colors of universal indicator at various pH values (see Figure 1 on p. 59).

Hazards:

Solutions of HCl can cause burns. Wear gloves and goggles.

Procedure:

1. Open each bottle and add 2-3 mL of the universal indicator. Recap and mix well. Note the color of the indicator and the approximate pH of the contents of each bottle.
2. Reopen and add 0.366 mL of 3 M HCl to each bottle. If using plastic droppers, try to add a little less than half a milliliter.
3. Recap each bottle and mix well. Note the color of indicator and approximate pH of the contents of each bottle. If the solution turns red, this indicates the water was unable to neutralize the added acid.
4. Continue to add 3 M HCl in increments of 0.366 mL to any bottle in which the indictor has not turned red in color.

Disposal:

Neutralize the solutions and place in a container designated for acid or base waste.

Discussion:

It is reasonable to wonder if the HCl in the human stomach can neutralize basic additives contained in alkaline water. The human stomach contains HCl at all times, and also produces HCl after a meal. These two factors should be considered to estimate if the human stomach produces enough HCl to neutralize the contents of various brands of alkaline water.

The contents of the fasting human stomach have an average pH of 1.5 and contain between 35 and 75 mL of fluid. Using the lower of these volumes, this works out to 1.1 mmol of H^+ contained in the fasting human stomach:

$$[H^+] = 10^{-pH} = 0.032\ M$$

$$\frac{0.032\ mol\ H^+}{L} x 0.035\ L = 0.0011\ mol\ H^+$$

In addition, the human stomach can produce up to 30 mmol H^+ per hour after a meal on average. This means the stomach can replenish the 1.1 mmol of H^+ present when fasting in only a little over two minutes:

$$1.1\ mmol\ H^+ x \frac{hr}{30\ mmol} x \frac{60\ min}{1\ hr} = 2.2\ minutes$$

In this experiment, additions of 0.366 mL of 3 M HCl are made to bottles of alkaline water. Each addition of 0.366 mL of 3 M HCl delivers approximately 1.1 mmol of H^+:

$$0.366 \times 10^{-3}\text{L} \times \frac{3\text{ mol HCl}}{\text{L}} = 0.0011\text{ mol H}^+$$

Generally speaking, one or two additions of 0.0011 mol of H^+ are sufficient to neutralize entire bottles of various brands of water, indicating that consuming alkaline water has no significant effect on the pH of the human body.

It is interesting to note the differences observed in this experiment between Fiji brand water (displays lower pH, but takes more acid to neutralize) and the Essentia brand water (displays higher pH, but takes less acid to neutralize). This comparison of the basic properties of Fiji versus Essentia water ends up providing a fantastic and simple demonstration of the difference between the ***strength*** and ***concentration*** of a base. The ingredients list of these waters indicates that Fiji water contains bicarbonate (HCO_3^-, $K_b = 5 \times 10^{-11}$), while Essentia water contains both bicarbonate and hydrogen phosphate (HPO_4^{2-}, $K_b = 1.6 \times 10^{-7}$). By comparing the K_b values for each ion, we see that HPO_4^{2-} is a much better base than HCO_3^- (over 3,000 times better!). Because of this, HPO_4^{2-} produces significantly more OH^- than HCO_3^-, due to its much larger K_b:

$$HCO_3^-(aq) + H_2O \leftrightarrow H_2CO_3(aq) + OH^-(aq) \quad (1)$$

$$HPO_4^{2-}(aq) + H_2O \leftrightarrow H_2PO_4^-(aq) + OH^-(aq) \quad (2)$$

So even though Fiji contains a greater **quantity** of basic ions capable of reacting with H^+, these ions are **weaker bases** and therefore ionize less extensively, producing less OH^-.

A titration of the number of protons required to completely acidify each brand of water may be done. A 700 mL bottle of Essentia brand water only requires about 0.57 mmol of H^+, whereas a 700 mL bottle of Fiji water requires about 1.8 mmol of H^+. Interestingly, the number of protons required to acidify the bottle of Fiji water matches quite well with the concentration of HCO_3^- listed on its ingredients label (152 mg L^{-1} HCO_3^-):

$$\frac{0.0018 \text{ mol } H^+}{0.700 \text{ L}} \times \frac{1 \text{ mol } HCO_3^-}{1 \text{ mol } H^+} \times \frac{61{,}000 \text{ mg } HCO_3^-}{1 \text{ mol } HCO_3^-} = 157 \text{ mg } HCO_3^- \text{ L}^{-1}$$

REFERENCE

1. Kuntzleman, T. (2022). Can alkaline water change body pH? *Chemical Education Xchange*. https://www.chemedx.org/blog/can-alkaline-water-change-body-ph

38

What Is a Buffer?

TOPICS: buffers, conjugate acids and bases

Materials:

- Dropper bottle filled 10% by weight HCl.
- Dropper bottle filled with 10% by weight NaOH.
- Deionized water.
- Two large beakers.
- A universal indicator.
- Alka-Seltzer tablet.
- Magnetic stir bars.
- Magnetic stir plates.
- A diagram for the audience to see the various colors of the universal indicator at different pH values (see Figure 1 on p. 59).

Hazards:

Solutions of HCl and NaOH can cause burns. Wear gloves and goggles.

Procedure:

1. Fill both beakers three-fourths full with water. Add a magnetic stir bar to each beaker, place each beaker on a stir plate, and get the water in each beaker stirring.
2. Add the universal indicator to each beaker and observe the yellow color, indicating slightly acidic pH due to dissolved CO_2.
3. Add an Alka-Seltzer tablet to one of the beakers. The other beaker is considered the control beaker. The solution temporarily turns red due to dissolved aspirin (a weak acid). However, the color in this beaker soon fades back the initial yellow color as bicarbonate reacts with aspirin and citric acid to buffer the solution.
4. Add two to three drops of HCl solution to both beakers. The beaker containing the Alka-Seltzer tablet shows almost no visible color change. On the other hand, in the control beaker, which contains no buffer while a dramatic change to red is observed.
5. Add five to seven drops of NaOH solution to both beakers. The beaker containing the Alka-Seltzer tablet shows almost no visible color change, while a dramatic change to red is observed in the control beaker.

Disposal:

Place solutions in the container indicated for aqueous base waste.

Discussion:

The color of the universal indicator in deionized water is yellow, indicating that water is slightly acidic. This result surprises many observers and therefore deserves some comment. Any water that is in contact with the atmosphere will contain dissolved CO_2:

$$CO_2(g) \rightarrow CO_2(aq) \quad (1)$$

Because CO_2 is a weak acid in water, this imparts a yellow color to the universal indicator:

$$CO_2(aq) + H_2O(l) \leftrightarrow H^+(aq) + HCO_3^-(aq) \quad (2)$$

Upon initial addition of the Alka-Seltzer tablet, the universal indicator turns red in color. This is likely due to aspirin (180 g mol^{-1}), a weak monoprotic acid, dissolving into the water (Figure 1). However, the Alka-Seltzer tablet also contains sodium bicarbonate (84 g mol^{-1}) and triprotic citric acid (192 g mol^{-1}). The sodium bicarbonate reacts with the citric acid and aspirin according to:

$$3\ NaHCO_3(aq) + H_3C_6H_5O_7(aq) \rightarrow Na_3C_6H_5O_7(aq) + 3\ H_2O(l) + 3\ CO_2(g) \qquad (3)$$

$$NaHCO_3(aq) + HC_9H_7O_4(aq) \rightarrow NaC_9H_7O_4(aq) + H_2O(l) + CO_2(g) \qquad (4)$$

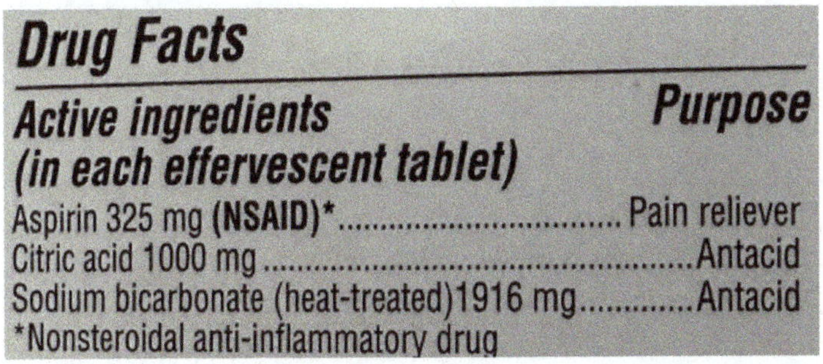

FIGURE 1: Active ingredients in an Alka-Seltzer tablet.

Based on the information in Figure 1, it can be calculated that an Alka-Seltzer tablet contains 17.4 mmol of titratable protons, and that an excess of sodium bicarbonate (22.8 mmol) is available to react with these protons. Thus, it can be estimated that 17.4 mmol of CO_2 is produced in this reaction, and 5.4 mmol of sodium bicarbonate remains. Certainly, much of this CO_2 escapes as a gas. The dissolved CO_2 that remains reacts reversibly with water and establishes an equilibrium with bicarbonate ions (HCO_3^-). The CO_2/HCO_3^- system behaves as a conjugate acid–base pair, producing a buffering effect. CO_2 acts as a Lewis acid, accepting electrons in this equilibrium, and the pK_a of the system (~6.35) explains why the buffered solution has a pH near this value, consistent with the yellow color observed.

Unsurprisingly, addition of strong acid or strong base to the control beaker causes drastic changes in pH (and therefore in color). However, almost no color change is observed in the buffered solution.

39

Seashell Chemistry 2

TOPICS: Le Chatelier's principle, chemical thermodynamics, Gibbs energy, K_{sp}, climate change

Materials:

- 500 mL of limewater (to prepare, boil 1 L of distilled water, cover to prevent exposure to air and allow to cool). Once cooled, add solid calcium hydroxide and shake well. Allow the calcium hydroxide to settle out, and then filter the limewater into a container. Seal well from air.
- 2 × 250 mL beakers.
- Straw.
- Dry ice.

Hazards:

Limewater is very basic. Wear gloves and goggles when handling. Dry ice is cold enough to cause burns Use tongs to handle it.

Procedure:

1. Divide the limewater equally into both beakers.
2. Add dry ice to the first beaker.
3. At the same time, begin blowing through a straw into the second beaker.
4. Notice that in both beakers, a cloudy-white precipitate of $CaCO_3$ forms. With prolonged bubbling of dry ice through the first beaker, the cloudy solid ends up dissolving.

Disposal:

Place solutions in the container indicated for aqueous base waste.

Discussion:

In this demonstration, CO_2 is bubbled into a saturated solution of calcium hydroxide (also known as limewater).[1-2] The CO_2 is introduced by addition of dry ice (solid carbon dioxide) or via exhaled breath (which contains 4% CO_2).[3] When exposed to CO_2, limewater turns milky white from $CaCO_3$ production through the following reactions:[1-2, 4-8]

$$CO_2(g) \leftrightarrow CO_2(aq) \quad (1)$$

$$CO_2(aq) + H_2O(l) \leftrightarrow H^+(aq) + HCO_3^-(aq) \quad (2)$$

$$Ca^{2+}(aq) + 2\,HCO_3^-(aq) \leftrightarrow CaCO_3(s) + H_2O(l) + CO_2(aq) \quad (3)$$

These reactions are instrumental in the formation of $CaCO_3$ in many marine organisms.[4-8] Curiously, if one adds dry ice (solid CO_2) to limewater, the $CaCO_3(s)$ that forms at first ends up dissolving as the CO_2 from the dry ice continues to bubble through the mixture. This effect is not observed when blowing exhaled breath into limewater.

Notice that the system of Equations 1-3 can be simplified by adding them together (Equation 1 + twice Equation 2 + Equation 3):

$$Ca^{2+}(aq) + CO_2(g) + H_2O(l) \leftrightarrow CaCO_3(s) + 2\,H^+(aq) \tag{4}$$

Although Equation 4 differs from standard presentations,[4–8] it is a useful composite reaction that is helpful in explaining how the limewater alternately forms or dissolves $CaCO_3$, depending upon conditions. Notice that CO_2 and H^+ impact which way Equation 4 will shift via the principle of Le Chatelier. Increased CO_2 pressure shifts the reaction toward formation of $CaCO_3$; in contrast, increased H^+ favors its dissolution. However, dissolved CO_2 concentration increases with CO_2 pressure (Equation 1), and increased $CO_2(aq)$ increases H^+ concentration (Equation 2). As a result, the very CO_2 that initially favors $CaCO_3$ formation can lead to its dissolution as the solution becomes more acidic.

Equation 4 can be analyzed quantitatively to provide further understanding of how the reaction behaves upon addition of dry ice to limewater. Using Gibbs energies of formation (Table 1), we calculate the Gibbs energy of the reaction outlined in Equation 4 under standard conditions, ΔG^o_{rxn}, to be +57 kJ mol^{-1}. The positive value of ΔG^o_{rxn} assures us that formation of $CaCO_3$ is not spontaneous under standard conditions.

Table 1: Standard Gibbs Energies of formation of substances in Equation 4.[9]

Substance	**ΔG_f^o (kJ mol^{-1})**
$Ca^{2+}(aq)$	−554
$CO_2(g)$	−395
$H_2O(l)$	−237
$CaCO_3(s)$	−1129
$H^+(aq)$	0

However, this demonstration is not carried out under standard conditions. For the Gibbs energy of the reaction under the conditions of this demonstration, the following is used:

$$\Delta G_{rxn} = \Delta G^o_{rxn} + RT\,lnQ \tag{5}$$

where R is the gas constant (8.314 J mol^{-1} K^{-1}), T is temperature, and Q for Equation 5 is:

$$Q = \frac{[H^+]^2}{[Ca^{2+}]P_{CO_2}} \tag{6}$$

Together, Equations 5 and 6 yield the following equation which can be used to calculate the Gibbs energy of the reaction under the conditions of this demonstration:

$$\Delta G_{rxn} = \Delta G^{o}_{rxn} + RT\, ln \frac{[H^{+}]^{2}}{[Ca^{2+}]P_{CO_2}} \quad (7)$$

When the dry ice is initially placed in the limewater, the following may be used to estimate the concentrations of all species is $[H^{+}] = 5 \times 10^{-13}$ M, $[Ca^{2+}] = 0.0108$ M, and $P_{CO2} = 1$ bar (see reference 10). Insertion of these values, $T = 298$ K, and $\Delta G^{o}_{rxn} = +57$ kJ mol^{-1} into Equation 7 yields $\Delta G^{o}_{rxn} = -72$ kJ mol^{-1}. Because the value of ΔG^{o}_{rxn} is negative, Equation 4 is predicted to be spontaneous when dry ice is first added to limewater. This is consistent with the formation of solid $CaCO_3$ initially observed in this demonstration. After exposure of the limewater to dry ice bubbles for a few minutes, the pH drops. A good estimate for the $[H^{+}]$ at this point is 1.2×10^{-4} M (see reference 10). Inserting this value and other relevant parameters ($[Ca^{2+}] = 0.0108$ M, $P_{CO2} = 1$ bar, $T = 298$ K, and $\Delta G^{o} = +\,57$ kJ mol^{-1}) into Equation 7 gives $\Delta G^{o}_{rxn} = +23$ kJ mol^{-1}.

Thus, after prolonged bubbling, the Gibbs energy for reaction 4 becomes positive. This indicates the reaction no longer proceeds spontaneously in the forward direction Thus, the reaction becomes nonspontaneous as written, and the reverse process—the dissolution of $CaCO_3$—is favored.

This experiment is a striking illustration of an adverse effect of increased atmospheric CO_2. As humans continue to emit fossil fuels, atmospheric CO_2 rises. This causes the acidity of Earth's oceans to increase, which negatively affects aquatic organisms that depend upon $CaCO_3$.[4–8]

REFERENCES

1. Shakhashiri, B. Z. (1983). Chemical demonstrations: A handbook for teachers of chemistry (Vol. 1, pp. 327–377). University of Wisconsin Press.
2. Bell, J. A. (2000). Every year begins a millennium. *Journal of Chemical Education, 77,* 1098–1102.
3. Tsoukias, N. M., Tannous, Z., Wilson, A. F., & George, S. C. (1998). Single-exhalation profiles of NO and CO_2 in humans: Effect of dynamically changing flow rate. *Applied Physiology, 85,* 642–652.
4. Feely, R. A. et al. (2004). Impact of anthropogenic CO_2 on the $CaCO_3$ system in the oceans. *Science, 305,* 362–366.
5. Weston, R. E. Jr. (2000). Climate change and its effects on coral reefs. *Journal of Chemical Education, 77,* 1574–1577.
6. Buth, J. M. (2016). Ocean acidification: Investigation and presentation of the effects of elevated carbon dioxide levels on seawater chemistry and calcareous organisms. *Journal of Chemical Education, 93,* 718–721.
7. Silverstein, T. P. (2022). Rising atmospheric carbon dioxide could doom ocean corals and shellfish: Simple thermodynamic calculations show why. *Jorunal of Chemical Education, 99,* 2020–2025.
8. Bozlee, B. J., Janebo, M., & Jahn, G. (2008). A simplified model to predict the effect of increasing atmospheric CO_2 on carbonate chemistry in the ocean. *Journal of Chemical Education, 85,* 213–217.
9. ChemNetBase. (n.d.). https://hbcp.chemnetbase.com/faces/documents/05_04/05_04_0001.xhtml
10. Kuntzleman, T. (2023). Seashell chemistry. *Chemical Education Xchange.* https://www.chemedx.org/blog/seashell-chemistry

40

K_{sp} and Magnesium Hydroxide

TOPIC: K_{sp}

Materials:

- Milk of magnesia—contains $Mg(OH)_2(s)$.
- 500 mL Erlenmeyer flask.
- Deionized water.
- A universal indicator.

Hazards:

Wear safety goggles and gloves.

Procedure:

1. Fill the 500 mL flask with 250 mL of deionized water.
2. Add one dropperful of milk of magnesia to the flask, and swirl to evenly suspend the solid particles.
3. Note the milky-white appearance of the resulting mixture. Instruct students that the mixture appears milky because milk of magnesia contains $Mg(OH)_2$, which is insoluble. Thus, several tiny solid particles of solid are suspended in the mixture. These particles scatter light, causing the cloudy appearance.
4. Add one dropperful of the universal indicator to the mixture. Note the purple color, which indicates pH ~10+ (see Figure 1 on p. 59).
5. Point out to students that the high pH indicates some dissolution of $Mg(OH)_2$, even though it is mostly insoluble.
6. To make the demonstration quantitative, calculate the expected pH of the solution using K_{sp} for $Mg(OH)_2$ (see Discussion).

Disposal:

Transfer the mixture to a receptacle indicated for basic waste.

Discussion:

Early in the chemistry curriculum, students are taught solubility rules, which classify various ionic compounds based on solubility in water. Some ionic compounds are soluble and dissolve readily in water, while insoluble ionic compounds do not. In this demonstration milk of magnesia is mixed into water. The resulting mixture is cloudy white in appearance due to the presence of insoluble $Mg(OH)_2$ in the milk of magnesia. However, when the pH of the mixture is tested with the universal indicator, it is found to be quite alkaline, with a pH ~10. This suggests that the $Mg(OH)_2(s)$ does dissolve to some extent, releasing OH^- ions into solution:

$$Mg(OH)_2(s) \leftrightarrow Mg^{2+}(aq) + 2\ OH^-(aq) \quad (1)$$

The degree to which $Mg(OH)_2$ (or any other insoluble ionic compound) dissolves is made quantitative through the solubility product constant, or K_{sp}, which is simply the equilibrium constant for the dissolution of the solid. Therefore, for $Mg(OH)_2$:

$$K_{sp} = [Mg^{2+}][OH^-]^2 = 1.8 \times 10^{-11} \quad (2)$$

Equation 2 allows us to calculate the expected pH of the mixture of milk of magnesia in water. Setting up an initial-change-equilibrium (ICE) chart for Equation 1, we have:

$$Mg(OH)_2(s) \leftrightarrow Mg^{2+}(aq) + 2\ OH^-(aq)$$

	$Mg(OH)_2(s)$	$Mg^{2+}(aq)$	$2\ OH^-(aq)$
I	–	0	0
C	–	$+x$	$+2x$
E	–	x	$2x$

Entering the equilibrium concentrations of $[Mg^{2+}]$ and $[OH^-]$ in the ICE chart into Equation 2 yields $4x^3 = 1.8 \times 10^{-11}$. Solving this equation for x, we find $x = 1.65 \times 10^{-4}$. Inspection of the ICE chart shows that this value of x corresponds to the following concentrations in the mixture of milk of magnesia and water:

$$[Mg^{2+}] = x = 1.7 \times 10^{-4}\ M$$

$$[OH^-] = 2x = 3.3 \times 10^{-4}\ M$$

From the hydroxide concentration, the pOH can be found:

$$pOH = -\log[OH^-] = -\log(3.3 \times 10^{-4}) = 3.48$$

Using pH + pOH = 14, the mixture is found to have pH = 10.52, in harmony with the observation that the universal indicator turns violet when added to the mixture of milk of magnesia in water.

41

Bubble Buster

TOPICS: K_{sp}, reaction quotient, mathematical manipulations of equilibrium constants, general interest

Materials:

- Bar soap, such as Ivory, that contains sodium palmitate ($NaC_{16}H_{31}O_2$).
- Deionized water.
- Tap water.
- Evian bottled water.
- Three test tubes.
- Optional: straws and 3 × 400 mL beakers.

Hazards:

Wear safety goggles to avoid splashing soapy water into eyes.

Procedure:

1. Rinse each test tube several times with deionized water.
2. Fill the first test tube half full with deionized water, the second test tube half full with tap water, and the third test tube half full with Evian water.
3. Shave a few pieces of Ivory soap into the test tube with deionized water. Stopper the test tube and shake vigorously for 15 s.
4. Note the degree of sudsing produced.
5. Inspect the sides of the test tube for evidence of solid (scum) formation.
6. Repeat steps 1 to 5 for tap water and then for Evian water.

Note: For a slightly more dramatic effect, this experiment may be done by adding 100 mL of each water type to separate 400 mL beakers. After shaving a few pieces of Ivory soap into each beaker, blow into each mixture with a straw, and note the degree of bubbly foam produced in each case.

Disposal:

Wash mixtures down the drain. The mixtures contain soap and water.

Discussion:

Many soaps, including Ivory, are sodium salts of long-chain carboxylic acids. For example, sodium palmitate ($NaC_{16}H_{31}O_2$) is a slightly soluble ionic compound with $K_{sp} = 3.2 \times 10^{-4}$:

$$NaC_{16}H_{31}O_2(s) \rightarrow Na^+(aq) + C_{16}H_{31}O_2^-(aq) \quad K_{sp} = 3.2 \times 10^{-4} \qquad (1)$$

It is the palmitate ion ($C_{16}H_{31}O_2^-$) that does the work of the soap. Many dissolved metal ions with a charge greater than +1 react with palmitate ions to form a solid. For example, Cu^{2+}, Mg^{2+}, Ca^{2+}, Fe^{2+}, and Al^{3+} all react with palmitate. Consider the reaction between dissolved Ca^{2+} ions and the palmitate ion:

$$Ca^{2+}(aq) + 2\ C_{16}H_{31}O_2^{-}(aq) \rightarrow Ca(C_{16}H_{31}O_2)_2(s) \qquad K_{eq,1} = 1.5 \times 10^{12} \qquad (2)$$

The solid calcium palmitate formed in Equation 2 is an undesirable product known as "soap scum." Note that the equilibrium constant for Equation 2 was found by taking the inverse of K_{sp} for calcium palmitate (6.5×10^{-13}). Taking the sum of Equation 2 and twice Equation 1 yields Equation 3:

$$2\ NaC_{16}H_{31}O_2(s) + Ca^{2+}(aq) \leftrightarrow 2\ Na^{+}(aq) + Ca(C_{16}H_{31}O_2)_2(s)\ K_{eq,2} = K_{sp}{}^{2}K_{eq,1} = 1.5 \times 10^{5} \quad (3)$$

Notice that, in Equation 3 the net result is the removal of palmitate ions from solution. The removal of palmitate ions from solution makes it so the soap cannot do its work anymore, making it unable to clean or create suds. Evian water typically produces little or no sudsing due to soap scum formation. This result can be quantitatively predicted in the following way. First, we write Equation 2 in terms of K_{sp} for calcium palmitate:

$$Ca(C_{16}H_{31}O_2)_2(s) \rightarrow Ca^{2+}(aq) + 2\ C_{16}H_{31}O_2^{-}(aq) \qquad K_{sp} = 6.5 \times 10^{-13} \qquad (4)$$

Using Equation 4, we note that precipitation will be favored under conditions where:

$$Q = [Ca^{2+}][C_{16}H_{31}O_2^{-}]^2 > K_{sp} \qquad (5)$$

To see if this condition is satisfied in Evian water, we need to know the concentration of Ca^{2+} and palmitate ion in the water under the conditions of the experiment. Finding the Ca^{2+} concentration is easy enough: inspection of a bottle of Evian water indicates that the concentration of Ca^{2+} in Evian water is 80 mg L^{-1} (Figure 1). This works out to 0.002 M Ca^{2+}.

FIGURE 1: Label on a bottle of Evian water.

It's a little trickier to find the palmitate concentration. Note that this can be found using the K_{sp} for sodium palmitate (Equation 1).

$$[Na^+]\,[C_{16}H_{31}O_2^-] = K_{sp} = 3.2 \times 10^{-4}$$

Assuming $[Na^+] = [C_{16}H_{31}O_2^-]$ and allowing $[C_{16}H_{31}O_2^-] = x$ means we can write:

$$x^2 = 3.2 \times 10^{-4}$$

From this we find that $x = [C_{16}H_{31}O_2^-] = 0.018$ M. Substitution of this value and $[Ca^{2+}] = 0.002$ M into Equation 5 yields the result:

$$Q = [0.002][0.018]^2 = 6.5 \times 10^{-7}$$

This is a million times bigger than K_{sp} for calcium palmitate. Therefore, precipitation of scum is indeed predicted in Evian water, consistent with observations.

Hard water contains dissolved metal ions such as Mg^{2+}, Ca^{2+}, and Cu^{2+}, which react with palmitate to form soap scum. Water softeners replace these ions with Na^+, which does not form insoluble compounds with palmitate. As a result, soft water produces more suds and is often preferred for washing.

42

Colorful Copper Equilibria

TOPICS: Le Chatelier's Principle, effect of concentration and temperature on chemical equilibria

Materials:

- 100 mL of 1 M $CuSO_4$.
- 20 g of table salt.
- Three large test tubes.
- 4 × 250 mL beakers.
- Heating unit.
- Deionized water.
- Ice.

Hazards:

Solutions of copper (II) sulfate are corrosive. Wear safety goggles.

Procedure:

1. Prior to the demonstration, get about 150 mL of water heating in one of the 250 mL beakers. To a second 250 mL beaker, add ice water.
2. Add the 1 M $CuSO_4$ to a 250 mL beaker.
3. Display the solution to the students, noting the blue color.
4. Add 20 g of table salt to the solution. Swirl to dissolve. Note the change in color.
5. Split the solution of $CuSO_4$/NaCl into equal portions in the three large test tubes.
6. Place one of the test tubes in the 250 mL beaker containing hot water. Place the second test tube in the 250 mL beaker containing ice water. Place the last test tube in an empty 250 mL beaker.
7. Observe any color changes that occur, comparing the colors at each temperature.

Disposal:

Transfer the contents of the test tubes to a container marked for copper waste.

Discussion:

In this demonstration, a blue-colored solution of Cu^{2+} is observed to shift to green and back again depending upon various conditions. These observations can be related to the formation of the yellow-colored trichloro copper (II) complex in the reversible reaction between Cu^{2+} and Cl^-:[1–10]

$$Cu^{2+}(aq) + 3\,Cl^-(aq) \leftrightarrow CuCl_3^-(aq) \quad (1)$$

When table salt is added to the $Cu^{2+}(aq)$ solution, a color change to green is observed. This is because the salt increases the concentration of dissolved Cl^- in solution, which shifts Equation 1 to the right by the principle of Le Chatelier. The color of the solution does not turn completely yellow because blue Cu^{2+} and yellow $CuCl_3^-$ are simultaneously present in the mixture. If this resulting mixture is heated, the solution shifts to a greener color. On the other hand, when this resulting mixture is cooled, it shifts to it appear bluer. These temperature dependent color changes can be understood by noting that the enthalpy change the of reaction outlined in Equation 1 has been reported[1] to be endothermic with $\Delta H = +11$ kJ mol^{-1}. Indeed, the principle of Le Chatelier indicates that endothermic reactions are shifted to the right (in this case, more green) with increases in temperature and to the left (in this case, more blue) when the temperature is decreased.

REFERENCES

1. Ramette, R. W., & Fan, G. (1983). *Inorganic Chemistry, 22*, 3323–3326.
2. Helmholz, L., & Kruh, R. F. (1952). *Journal of American Chemical Society, 74*, 1176–1181.
3. Khan, M. A., & Schwing-Weill, M. J. (1976). Stability and electronic spectra of the copper(II) chloro complexes in aqueous solutions. *Inorganic Chemistry, 15*, 2202–2205.
4. Elleb, M., Meullemeestre, J., Schwing-Weill, M. J., & Vierling, F. (1980). Stability, electronic spectra, and structure of the copper(II) chloride complexes in N,N-dimethylformamide. *Inorganic Chemistry, 19*, 2699–2704.
5. Khan, M. A., Meullemeestre, J., Schwing-Weill, M. J., & Vierling, F. (1989). Detailed spectrophotometric study of the copper(II) halides in anhydrous methanol. *Inorganic Chemistry, 28*, 3306–3309.
6. Benghanem, F., Khan, M. A., Meullemeestre, J., & Vierling, F. (1991). Halogenocomplexes of copper(II) in anhydrous propan-2-ol. *Polyhedron, 10*, 2529–2533.
7. Katzin, L. I. (1962). Ionization differences between coordination states of a cation. Octahedral–tetrahedral equilibrium of transition-element chlorides in dimethylformamide. *Journal of Chemical Physics, 36*, 3034–3041.
8. De Vreese, P., Brooks, N. R., Van Hecke, K., Van Meervelt, L., Matthijs, E., Binnemans, K., and Van Deun, R. (2012). Speciation of copper(II) complexes in an ionic liquid based on choline chloride and in choline chloride/water mixtures. *Inorganic Chemistry, 51*, 4972–4981.
9. Ohtaki, H. (1987). Structural studies on solvation and complexation of metal ions in nonaqueous solutions. *Pure & Applied Chemistry, 59*, 1143–1150.
10. Kuntzleman, T. (2020). A multi-colored equilibrium experiment. *Chemical Education Xchange.* https://www.chemedx.org/blog/multi-colored-equilibrium-experiment

43

Formation of a Copper Complex

TOPIC: Complex ion formation

Materials:

- 100 mL of 0.01 M $CuSO_4$.
- Household ammonia.
- 250 mL Erlenmeyer flask.

Hazards:

Solutions of copper (II) sulfate and ammonia are corrosive. Wear safety goggles.

Procedure:

1. Add 100 mL of 0.010 M $CuSO_4$ to the Erlenmeyer flask. Note the pale blue color of the solution.

2. Add 50 to 100 mL of ammonia. Note the color change to deep blue.

Disposal:

Transfer the contents of the flask to a container marked for copper waste.

Discussion:

In this simple demonstration, the addition of a solution of ammonia to a solution of $Cu^{2+}(aq)$ causes a color change from pale blue to deep blue. This color change is due to the formation of the tetraamminecopper (II) ion:

$$Cu^{2+}(aq) + 4\,NH_3(aq) \leftrightarrow Cu(NH_3)_4^{2+}(aq) \qquad K_f = 1.1 \times 10^{13} \qquad (1)$$

The equilibrium constant for Equation 1, K_f, is:

$$K_f = \frac{[Cu(NH_3)_4^{2+}]_{eq}}{[Cu^{2+}]_{eq}[NH_3]_{eq}^4} \qquad (2)$$

where the eq subscripts represent equilibrium concentrations.

We will assume this demonstration used concentrations of 0.5 M NH_3 and 0.01 M Cu^{2+} to predict the equilibrium concentrations of all species in this experiment. To do so, we will set up an "initial, change, equilibrium" (ICE) chart. To make the calculations easier, however, we'll assume that upon mixing the 0.50 M NH_3 and 0.01 M Cu^{2+} solutions, all the Cu^{2+} is converted into $Cu(NH_3)_4^{2+}$. Thus, the initial concentration of $Cu(NH_3)_4^{2+}$ = 0.010 M and the initial concentration of Cu^{2+} = 0 M. To form the 0.010 M $Cu(NH_3)_4^{2+}$, 0.040 M of NH_3 must have reacted. Therefore, the initial concentration of $NH_3 = 0.50 - 4(0.010\ M) = 0.46\ M$. We substitute these values into the initial row of the ICE chart:

$$Cu^{2+}(aq) + 4\ NH_3(aq) \leftrightarrow Cu(NH_3)_4^{2+}(aq)$$

	$Cu^{2+}(aq)$	$4\ NH_3(aq)$	$Cu(NH_3)_4^{2+}(aq)$
Initial	0 M	0.46 M	0.010 M
Change	$+x$	$+4x$	$-x$
Equilibrium	x	$0.46 + 4x$	$0.010 - x$

Substitution of the equilibrium values found in the above chart into Equation 2 yields:

$$K_f = \frac{[0.010-x]}{[x][0.46+4x]^4} = 1.1 \times 10^{11} \quad (3)$$

Because the value of the equilibrium constant is so very high, we can assume that the reaction will not proceed very far in the reverse direction. Because of this, we can assume that x is very small compared to 0.010 on the numerator of Equation 3, while 4x is very small compared to 0.46 on the denominator of Equation 3. As such, Equation 3 is simplified to the following:

$$\frac{[0.010]}{[x][0.46]^4} = 1.1 \times 10^{11} \quad (4)$$

Solving Equation 4 for x, we find $x = 2.0 \times 10^{-12}$ M. Indeed, our assumptions that x is very small compared to 0.01, and $4x$ is very small to 0.46 were justified. Substituting this value of x into the bottom, equilibrium row of the ICE chart, we have the following equilibrium concentrations:

$$[Cu^{2+}]_{eq} = 2.0 \times 10^{-12}\ M,\ [NH_3]_{eq} = 0.46\ M, \text{ and } [Cu(NH_3)_4^{2+}]_{eq} = 0.010\ M.$$

These calculations predict that the reaction should essentially run to completion, consistent with what is observed in this experiment.

44

Play-Doh Battery

TOPIC: Electrochemical Cells

Materials:

- Play-Doh.
- U.S. pennies.
- Galvanized washers.
- Red LED.
- Voltmeter (optional).

Procedure:

1. Sandwich a small amount of Play-Doh between a penny and a washer to create a single cell.
2. Repeat step 1 to create three or four cells.
3. Stack the cells on top of each other. The copper penny of one cell should be connected to the zinc washer of an adjacent cell. This combination of cells is known as a battery of cells.

4. Connect the long wire of the LED to the copper end of the battery of cells. Simultaneously connect the short wire of the LED to the zinc end of the battery of cells. If the LED does not light, try reversing the connections.
5. If desired, use a voltmeter to measure the potential of one of these cells.

Discussion:

This simple demonstration introduces students to galvanic cells using simple household items such as Play-Doh. When several of these cells are connected in series, they supply a high enough electric potential to power a red LED. The chemical reaction that powers these cells has been described to be likely the following[1–2]:

$$2\ Zn(s) + O_2(g) + 4\ H^+(aq) \rightarrow 2\ Zn^{2+}(aq) + 2\ H_2O(l) \qquad (1)$$

In Equation 1, zinc is oxidized, and oxygen is reduced. In the cell used in this experiment, zinc on the galvanized washer serves as the anode, the penny serves as the cathode, the Play-Doh serves as the salt bridge, and the dissolved salt within the Play-Doh serves as the electrolyte.

The overall reaction in Equation 1 can be written as the sum of the following two half-reactions:

$$Zn(s) \leftrightarrow Zn^{2+}(aq) + 2\ e^- \qquad E^0 = +0.76\ V \qquad (2)$$

$$O_2(g) + 4\ H^+(aq) + 4e^- \leftrightarrow 2\ H_2O(l) \qquad E^0 = +1.23\ V \qquad (3)$$

Note that the sum of twice Equation 2 and Equation 3 yields Equation 1 and a cell potential $E^0 = +1.99$ V. However, in my experience the cell potential measured in this experiment with a voltmeter never measures higher than 1 V. More often than not, values of 0.5–0.8 V are observed. These observations can be made semiquantitative by assuming that the reaction at the cathode involves the reduction of copper:

$$Cu^{2+}(aq) + 2\ e^- \leftrightarrow Cu(s) \qquad E^0 = +0.34\ V \qquad (4)$$

The sum of Equations 2 and 4 yield the following overall reaction with a cell potential equal to +1.10 V, slightly more consistent with observations:

$$Cu^{2+}(aq) + Zn(s) \leftrightarrow Cu(s) + Zn^{2+}(aq) \qquad E^0 = +1.10\ V \qquad (5)$$

REFERENCES

1. O'Donoghue, J. et al. (2024). *Journal of Chemical Education, 101*, 5089–5096.
2. Clarke, T. B., Glasscott, M. W., & Dick, J. E. (2021). The role of oxygen in the voltaic pile. *Journal of Chemical Education, 98*, 2927–2936.

www.ingramcontent.com/pod-product-compliance
Lightning Source LLC
La Vergne TN
LVHW060450061126
851857LV00016B/1782

* 9 7 9 8 3 8 5 1 9 2 9 1 5 *